HOLT
PHYSICS

TECHNOLOGY-BASED LAB ACTIVITIES
CBL™ EXPERIMENTS

TEACHER'S EDITION

HOLT, RINEHART AND WINSTON

A Harcourt Classroom Education Company

Austin · New York · Orlando · Atlanta · San Francisco · Boston · Dallas · Toronto · London

Holt Physics

Technology-Based Lab Activities
CBL™ Experiments
Teacher's Edition

Contributing Writer

E. David Thielk
Science Educator
Peninsula College
Port Townsend, WA

Reviewers

Richard Sorensen
Vernier Software & Technology
Beaverton, OR

Martin Taylor
Sargent-Welch/CENCO Physics
Buffalo Grove, IL

Safety Reviewer

Gregory Puskar
Laboratory Manager
Department of Physics
West Virginia University
Morgantown, WV

Cover photo: © Lawrence Manning/CORBIS
Cover design: Jason Wilson

Printed in the United States of America

ISBN 0-03-057343-2

2 3 4 5 6 095 05 04 03 02

Contents

Introduction to the Technology-Based Lab Activities

Calculator-Based Laboratory™ (CBL™) Technology

Calculator-Based Laboratory™ (CBL™) technology from Texas Instruments, coupled with software and probes by Vernier Software and Technology, turns every student's graphing calculator into a powerful lab instrument.

The CBL is a separate unit that links to a graphing calculator with a black link cable. Various probes, such as motion detectors, force sensors, temperature probes, light sensors, or voltage probes, can be plugged into the CBL, allowing the calculator and the CBL to collect physical data with a high degree of precision. The calculator can then be used to analyze the data, plot graphs, and calculate final results.

CBL activities are safe, easy to set up and clean up, and often less expensive than traditional lab experiments.

The activities in this booklet are designed for the original CBL units from Texas Instruments and for TI-83 Plus graphing calculators.

---

internet connect

GO TO: go.hrw.com
KEYWORD: HF2 TECHLABS
These activities may also be performed with TI CBL 2 or Vernier LabPro systems. Instructions are available at this Web site.

Tried-and-True Procedures

The Technology-Based Lab Activities are adapted from experiments developed by Vernier Software & Technology, a leading manufacturer of software and probes for CBL systems. The procedures for their manuals have been tried repeatedly with excellent results. Every activity in this booklet has been reviewed and approved by Vernier Software & Technology.

Each procedure gives step-by-step instructions for setting up, calibrating, and collecting data with CBL systems and sensors. Keystrokes on the calculator are shown in calculator fonts, and text that will appear on the calculator screen is shown as it will appear to students.

A Scenario-Based Approach

Scientists and engineers often go to the laboratory to develop models of physical situations found in the world outside the lab. Each of the activities in this booklet presents a real-world scenario or problem for students to model in the laboratory. As students work on these activities, they will develop a deeper understanding of how the concepts presented in their physics textbook relate to everyday physical phenomena, and they will use their understanding of physics to solve problems like those faced by physicists and engineers everyday. The modeling and problem-solving skills students will learn can be applied to many different situations they may encounter in their future personal and professional lives.

Modeling the Physical World

Several different types of models may be used in a physics lab. A *physical model* is a physical representation of a real physical system, but the model is simplified and is often a small-scale representation. A *mathematical model* is an equation or set of equations that can be used to represent a physical phenomenon or relationship. A *graphical model* is a graph or set of graphs that can be used to represent the relationships between physical quantities. Students will encounter all three of these types of models as they work through the activities in this booklet.

Each activity contains a Developing the Model section that helps students explore the connections between theoretical concepts and real-world scenarios. You may assign these exercises as homework to be done by students before they come to the lab, or you may work through them with the class.

Students test their models by collecting and analyzing data with CBL systems and graphing calculators. Steps in the procedure may be repeated as the model is refined and improved. Once students have established a good working model, they use the model to answer questions or solve problems in the original real-world scenario. Students also evaluate the accuracy and precision of their models and examine underlying assumptions.

About the Teacher's Notes

Teacher's notes for each lab are found in the first half of this booklet. The entire student booklet is included in the second half. The teacher's notes for each lab contain the following items.

- **Materials lists, materials preparation, and classroom organization** sections help you plan and prepare the lab before students arrive.

- **Techniques to Demonstrate** give tips on tricky techniques and concepts to explain to students before they begin.

- **Checkpoints** point out key steps in the procedure where students may need help. You should check on student progress as they reach these critical points.

- **Sample data and sample graphs** are provided for every lab. Sample data may be used to demonstrate calculations, and sample graphs help you evaluate student results by providing examples.

- **Answers to questions** from the Developing the Model, Analysis, Conclusions, and Extensions sections help you evaluate lab reports. When appropriate, sample answers are presented as calculator screen shots or graphs.

This Teacher's Edition also contains a **master materials list** and comprehensive **teacher guidelines for laboratory safety** to help you keep your laboratory safe and fully equipped.

Using CBL systems

To download required software for graphing calculators, you will need a computer with an Internet connection, a TI-GRAPH LINK cable, and a TI-83 Plus or compatible calculator. Visit the go.hrw.com Web site, and type "HF2 CALC" at the keyword prompt.

↗ internetconnect

go. hrw .com
GO TO: go.hrw.com
KEYWORD: HF2 CALC
This Web site contains links for downloading programs and applications you and your students will need for the Technology-Based Lab Activities.

---

1. If your computer does not already have TI-GRAPH LINK software installed, click Install TI-GRAPH LINK and follow the links for downloading and installing TI-GRAPH LINK from the Texas Instruments Web site.

2. Click Download the PHYSICS App and follow the links for downloading the PHYSICS application from the Vernier Software & Technology Web site. You will also need to follow the instructions for your TI GRAPH-LINK to load the application onto your calculator. Once the PHYSICS application is loaded onto your calculator, it should appear in the APPS menu.

3. If you need more instructions on using the CBL system, click CBL Made Easy to view a tutorial from Vernier Software & Technology.

Every calculator that will be used with a CBL unit needs to have the PHYSICS application loaded. Students are given instructions for downloading and installing the PHYSICS application, but not all students will have TI-GRAPH LINK cables, and some may not have access to a computer or the Internet. The PHYSICS application can also be transferred directly between calculators using a unit-to-unit cable. Refer to the TI Web site or your calculator's user's manual for instructions.

Troubleshooting

- Calculator and CBL instructions in the *Holt Physics* program are written for the TI-83 Plus, for the original CBL from Texas Instruments, and for probes from Vernier Software & Technology. If you use other hardware, some of the programs and instructions may not work exactly as described.

- If you have problems loading programs or applications onto your calculator, you may need to clear programs or other data from your calculator's memory.

- Always make sure that you are downloading correct versions of the software. TI-GRAPH LINK and the PHYSICS application both have different versions for different types of computers as well as for different calculators.

Getting Help

If you need additional help, both Texas Instruments and Vernier Software & Technology can provide information, customer service, and technical support.

Contacting Texas Instruments
email: ti-cares@ti.com
address: Texas Instruments
P.O. Box 650311, MS 3962
Dallas, TX 75265
Customer Support phone: 1-800-TI-CARES
Technical Support phone: (972) 917-8324

Contacting Vernier Software & Technology
email: info@vernier.com
address: Vernier Software & Technology
13979 Millikan Way
Beaverton, OR 97005
phone: (503) 277-2299

Making your laboratory a safe place to work and learn

Safety should be an integral part of any activity, and especially in the laboratory. Unfortunately, safety tends to be like the speed limit—we are aware of it, but we seldom heed it. We mistakenly believe that nothing bad will happen in our activities. However, no activity is completely free of risk, and it is rather late to worry about safety after an accident. By reviewing and evaluating safety issues before conducting any laboratory activities, teachers can minimize risk to students and prevent most accidents.

Addressing these concerns takes time and effort. Safety must be an integral part of the instructional process, not only to meet legal, ethical, and moral obligations, but also to promote safe behaviors and attitudes that students can take to their homes and workplaces in the future.

Concern for safety must begin prior to any classroom activity and before students enter the lab. A careful review of the facilities should be a basic part of preparing for each school term. The first steps you should take are to investigate the physical environment, identify any safety risks, and inspect your work areas for compliance to safety regulations.

The review of the lab should be thorough, and all safety issues must be addressed immediately. Keep a file of your review, and add to the list each year; this will allow you to continue to raise the standard of safety in your classroom. Do not perform any activities in a classroom or laboratory unless you can verify that the laboratory is safe and you are prepared to respond to emergency situations.

Monitoring activities for safety

Once the laboratory is reviewed and all safety issues are corrected, turn your attention to the activities that will be performed in class. Students should not be allowed to work in the lab until procedures have been reviewed and approved from a safety standpoint. Every teacher should review all experiments, demonstrations, and activities for safety concerns before presenting them to the class. The goal of this review is to identify and eliminate potential safety hazards. In this way, you can maximize learning while minimizing risk and can be prepared in case a safety issue does arise during the activity.

All reviews should be carried out in the following four stages.

1. Identify the risks.

Before introducing any activity to the class, analyze it and ask yourself what could possibly go wrong. Try to consider all possibilities, no matter how unlikely they may seem. Carefully review the list of materials called for to make sure they can be used safely in your classroom. Inspect the apparatus in your classroom to make sure it is in good working order. Read the procedures to make sure they are safe. Record any hazards or concerns you identify in a safety log.

2. Evaluate the risks.

The focus of this step is to minimize the risks you identified in the last step without sacrificing learning. Weigh the pedagogical value of the activity against the safety risks. Remember that no activity you can perform in the lab is worth the risk of injury. Thus, extremely hazardous activities or those which violate your school's policies must be eliminated. For activities that present smaller risks, analyze each risk carefully to determine its likelihood. Using your best judgment and all available information, estimate the probability of occurrence and possible consequences for each risk. If the pedagogical value of the activity does not outweigh the risks, the activity must be eliminated.

3. Select controls to address risks.

Even low-risk activities require controls to minimize or eliminate the risks. Be careful that in devising controls you do not substitute an equally or more hazardous alternative. If no reasonable controls can be devised, eliminate the activity. Some control methods include the following.

- Explicit verbal and written warnings may be added, signs may be posted, and labels may be attached.

- Apparatus may be rebuilt, may be relocated, may have parts replaced, or may be replaced entirely by safer alternatives.

- Risky procedures may be changed or eliminated.

- Student activities can be performed as teacher demonstrations.

4. Implement and review selected controls.

Controls do not help if they are forgotten or not enforced. The implementation and review of controls should be as systematic and thorough as the initial analysis of safety concerns in the lab and laboratory activities. This constant, ongoing procedure should take precedence over other concerns.

The following list describes several possible safety hazards and controls that can be implemented to resolve them. The list is not intended to be complete, but it can be used as a starting point to identify some of the known hazards in your laboratory.

Identified safety risk	Preventative control
Facilities and equipment	
Lab tables are in disrepair, the room is poorly lighted, and faucets and electrical outlets do not work or are difficult to use because of their locations.	Work surfaces should be level and stable. Adequate lighting should be available. Water supplies, drains, and electrical outlets should be in good working order. Any equipment in a dangerous location should be relocated or rendered inoperable.
Wiring, plumbing, and air circulation systems do not work or do not meet current specifications.	Specifications should be kept on file. Conduct a periodic review of all equipment, and document compliance. Damaged fixtures must be labeled and repaired as soon as possible.
Labs are conducted in multipurpose rooms, and equipment from other courses remains accessible.	Only the items necessary for a given activity should be available to students. All equipment should be locked away when not in use.
Students are permitted to enter or work in the lab without teacher supervision.	Lock all laboratory rooms when a teacher is not present. Supervising teachers must be trained in lab safety and emergency procedures.

Identified safety risk	Preventative control
Safety equipment and emergency procedures	
Fire and other emergency drills are infrequent, and no records or measurements are made of the results of the drills.	Implement regular fire and other emergency drills. Be sure that plans include alternate routes. Don't wait until an emergency to find the flaws in your plans.
Emergency evacuation plans do not include instructions for securing the lab in the event of an evacuation during a lab activity.	Plan actions in case of an emergency: establish what devices should be turned off, which escape route to use, and where to meet outside the building.
Fire extinguishers are in inconvenient locations, not on the escape route.	Place fire extinguishers near escape routes so that they will be useful during an emergency.
Fire extinguishers are not maintained regularly. Teachers are not trained to use extinguishers.	Document the regular maintenance of extinguishers. Train supervisory personnel in the proper use of extinguishers. Instruct students not to use an extinguisher but to call for a teacher.
Teachers in labs and neighboring classrooms are not trained in CPR or first aid.	Teachers should receive training from the local chapter of the American Red Cross. Certifications should be kept current with frequent refresher courses.
Teachers are not aware of their legal responsibilities in case of an injury or accident.	Review your faculty handbook for your responsibilities regarding safety in the classroom and laboratory. Contact the legal counsel for your school district to find out the extent of their support and any rules, regulations, or procedures you must follow. Train all supervisory teachers in the appropriate procedures.
Emergency procedures are not posted. Emergency numbers are kept only at the main switchboard. Instructions are given verbally, at the beginning of each school year, and never reviewed.	Emergency procedures should be posted at all exits and near all safety equipment. Emergency numbers should be posted at all phones, and a script should be provided for the caller to use. Emergency procedures must be reviewed periodically, and students should be reminded of them at the beginning of each activity.
Spills are handled on a case-by-case basis and are cleaned up with whatever materials happen to be on hand.	Have the appropriate equipment and materials available for cleaning up; replace it before expiration dates. Make sure students know to alert you to spilled chemicals, blood, and broken glass.

Identified safety risk	Preventative control
Work habits and environment	
Safety wear is only used for activities involving chemicals or hot plates.	Aprons and goggles should be worn in the lab at all times. Long hair, loose clothing, and loose jewelry should be secured.
No dress code is established for the laboratory; students are allowed to wear sandals or open-toed shoes.	Open-toed shoes should never be worn in the laboratory. Do not allow any footwear in the lab that does not cover feet completely.
Students are required to wear safety gear but teachers and visitors are not.	Always wear safety gear in the lab. Keep extra equipment on hand for visitors.
Safety is emphasized at the beginning of the term but is not mentioned later in the year.	Safety must be the first priority in all lab work. Students should be warned of risks and instructed in emergency procedures for each activity.
No assessment is made of students' knowledge and attitudes regarding safety.	Conduct frequent safety quizzes. Only students with perfect scores should be allowed to work in the lab.
Safety inspections are conducted irregularly and are not documented. Teachers and administrators are unaware of what documentation will be necessary in case of a lawsuit.	Safety reviews should be frequent and regular. All reviews should be documented, and improvements must be implemented immediately. Contact legal counsel for your district to make sure your procedures will protect you in case of a lawsuit.

Technology-Based Lab Activities
Master Materials List

CBL Equipment, Sensors, and Probes (available from Vernier Software & Technology or CENCO/Sargent-Welch)

Item	Quantity per group	Vernier Software Item Number	CENCO/Sargent-Welch Item Number	Labs
Calculator-Based Laboratory (CBL) system	1	CBL	WLS-13270	ALL
TI-83 Plus graphing calculator	1	TI-83PL	WLS-13283-P	ALL
PHYSICS application loaded in calculator	1	(See p. vii for instructions)		ALL
DIN adapter for CBL	1	DIN-BTA	WLS-13270-X	2, 4, 6, 7, 12, 19, 20, 21
Accelerometer, low-g or 5-g	1	LGA-BTA	CP33665-00	2A, 7
Calculator-Based Ranger (CBR) system (option to replace CBL and motion detector)	1	CBR	WLS-13271	1, 4, 5
Current and voltage probe set (includes dual-channel amplifier)	1	CV-BTA	CP33672-00	19, 20
Force sensor	1	DFS-BTA	CP33663-00	4, 6
Light sensor	1	LS-BTA or included w/CBL	Included in WLS-13270	14
Magnetic field sensor	1	MG-BTA	CP33669-00	21
Microphone	1	MCA-BTA	WLS-13270-E	13
Motion detector	1	MD-BTD	WLS-13270-A	1, 4, 5, 6
Photogate	1	VPG-BTD	CP36802-00	2B, 3, 12
Photogate adapter for CBL, dual	1	CBL-2P		3
Photogate adapter for CBL, single	1	CBL-P		2A, 2B
Picket fence	1	PF	CP32723-00	2B
Temperature sensor	1	TMP-BTA or included w/CBL	Included in WLS-13270	10
Voltage probe	1	VP-BTA or included w/CBL	Included in WLS-13270	18, 21

Other Equipment and Supplies (available from CENCO/Sargent-Welch)

Item	Quantity per group	CENCO/Sargent-Welch Item Number	Labs
Alligator clip cords, set of 10	1	WLA5419A	18, 20
Bottle with screw cap, plastic, 60 mL	1	WLS-8416-B	2A
Buret clamp, plastic-coated jaws	1	WL4901	19
Cardboard screen	1	WL3612	14
Circular spirit level	1	WLS-44360	7
Contact key	1	CP33638-00	20, 21
Cord, nylon*	1 spool	CP88066-00	2, 3, 4, 6, 12
DC Motor, 3V	1	WL2454C	21
Dynamics cart	1	CP72700-84	6
Elastic cord (surgical tubing), 1 m*	1	WLS-73675	2A

Technology-Based Lab Activities Master Materials List continued

Other Equipment and Supplies (available from CENCO/Sargent-Welch)

Item	Quantity per group	CENCO/Sargent-Welch Item Number	Labs
Electric buzzer	1	CP84020-00	21
Electronic balance, 200 g capacity × 0.1 g	1	WLS-2646-54A	4B, 10
Electronic balance, 2000 g capacity × 1 g	1	WLS-2646-59A	4A
Enameled copper magnet wire, 26 AWG*	1	WLS-85135-40J	21
Energy transfer kit	1	CP72700-72	19
Extension ring, 3″ diameter	1	WLS-73055-B	2B
Laminated polarizing filters	1 pair	CP87618-01	14
Lamp board, single	1	CP32144-01	21
Magnetic compass	1	CP78430-04	21
Masking tape, roll 3/4″ × 150′*	1	WLS-44182-80	1, 3, 7
Meterstick	1	WLS-44685	1, 2, 3, 7, 12
Pendulum clamp	1	WL0828	12
Plumb bob, nickel-plated steel	1	WLS-70200	3
Power supply, 12 V AC/DC compact	1	WLS-30970-20	19, 20, 21
Protractor, plastic, 6″ diameter	1	WLS-44785-20	12, 14
Right angle clamp for 1/2″ rods	2	WLS-19320	2B, 3, 6
Single-pole, double-throw switch	1	WLS-31155-B	21
Slotted mass, 5 g	1	WLS-4320-20C	19
Slotted mass, 10 g	2	WLS-4320-20D	19
Slotted mass, 100 g	3	WLS-4320-20G	12
Slotted mass, 1/2 kg	1	WLS-4321-20A	4A, 6
Slotted mass, 1 kg	1	WLS-4321-20B	4A
Steel ball, 19 mm diameter, solid	1	WLS-4481-D	3
Straight filament lamp, 40 W, with base	1	CP32944-00	14
Support base, rectangular, with 24″ rod	2	WLS-78306-C	3, 6, 12, 14, 19
Support rod, steel, 1/2″ × 24″	1	WLS-78379-E	2A, 2B
Table clamp	1	WLS-78430-20	2A, 2B
Threaded rod, 3/8″ × 8″	1	WLS-78379-A	2B
Transparent tape, roll 1/2″ × 150′*	1	WLS-44178	2B, 14
Tuning fork, 426.7 Hz (A)	1	WL3223	13
Volumetric transfer pipets, 7 mL, polyethylene	1	WLS-69684-30B	10

Equipment and Supplies Available Locally

Item	Quantity per group	Labs
35 mm film canister with lid	1	10
Balloon, baseball-sized	1	2B
Baseball	1	2B
Battery, 9 V, with battery clip*	1	18
Bungee cord	1	2A
Capacitor, 10 μF, nonpolarized	1	18

Item	Quantity per group	Labs
Carbon paper*	1 sheet	3
Cardboard, 2′ × 2′ piece	1	1, 3
Coffee filters, basket-style	5	4B
Composition roof shingle	1	4
Cushion, pillow, or padded catch box	1	2B
Elastic cord	1	6
Graph paper*	several sheets	ALL
Gravel	8 oz	2A
Hot water*	4 oz	10
Insulating fabrics (wool, fleece, polypropylene, etc.	2	10
Newspaper, 1 sheet*	1	2B
Packing tape, roll*	1	4A
Pair of polarized sunglasses	1	14
Plastic ball	1	2B
Ramp	1	3
Resistor, 1 Ω	1	21
Resistor, 22 Ω	1	18
Resistor, 47 Ω	1	18
Resistor, 50 Ω	2	20
Resistor, 68 Ω	2	20
Rubber band, long	1	2A
Rubber bands	several	2A, 6, 10
Shoes (not from the same pair)	2	4A
Small square box or square or circular frame	1	21
Stringed musical instrument (or any tunable instrument	1	13
Toothpick or wooden skewer	1	14
Turntable with 33⅓, 45, 78 rpm settings	1	7
Volleyball, soccer ball, or other similar ball	1	5
Waistband elastic	1	2A
Wire basket, sturdy	1	5

*An asterisk indicates that the item is consumable. All other items may be reused.

Technology Lab

Chapter 1

Graph Matching

OBJECTIVES

In this lab, students will:

- **Model** the motion of a robotic arm by moving an object in front of a motion detector and generating a distance-time graph.

- **Predict, sketch, and test** distance-time graphs that may be used as programming instructions for a robotic arm.

- **Predict, sketch, and test** velocity-time graphs that may be used as programming instructions for a robotic arm.

- **Evaluate** the use of graphs to provide instructions or analysis tools for robotic arms.

PLANNING

Recommended Time

1 lab period

Materials

[for each lab group]

✔ graphing calculator with link cable

✔ CBL system

✔ PHYSICS application loaded in calculator

✔ Vernier motion detector

✔ 2 ft. × 2 ft. piece of cardboard

✔ meterstick

✔ masking tape

Classroom Organization

- This activity should be performed in groups of two or more.

- If possible, the equipment should be set up so that each lab group has an open space of several meters in length. You may choose to complete this activity outdoors or in a gymnasium.

Materials Preparation

You may want to set up lab stations beforehand. If you have more than one lab class, the equipment can be left standing from one class to the next.

It is helpful to use masking tape and place marks on the floor at 1 m intervals. If it is not practical to place masking tape on the floor, omit these instructions from the lab handouts. Carefully lay out the path of motion for each station. Make sure paths are free of obstructions. Paths should not overlap or run too close to one another.

The TI CBR™ (Calculator-Based Ranger™) can be used to replace both the Vernier motion detector and the CBL system. Use a black link cable to connect the CBR directly to the calculator, and use the PHYSICS application as directed. If you have CBRs available, direct connection is convenient because fewer cables and pieces of equipment are required.

🖳 internet**connect**
GO TO: go.hrw.com **KEYWORD:** HF2 TECHLABS This lab may also be performed with TI CBL 2 or Vernier LabPro systems. Instructions are available at this Web site.

Techniques to Demonstrate

This experiment may be the first time your students use the motion detector. Providing some coaching on its use now will save time later in the year, because the motion detector is used in several experiments. Following are some hints for effective use of the motion detector.

- When using the motion detector, the detector emits ultrasound in a cone about 30° wide. Anything within this cone can cause a reflection of the ultrasound and possibly an accidental measurement. A common problem in using the motion detector is getting unintentional reflections from a desk, chair, or table in the room or from a wall or ceiling behind the target.

- The motion detector does not properly detect objects closer than 0.4 m. The maximum range is about 6 m, but stray objects in the wide detection cone can be a problem at this distance.

- Sometimes a target may not supply a strong reflection of the ultrasound. For example, if the target is a person wearing a bulky sweater, the resulting graph may be inconsistent.

- If graphs, especially velocity graphs, are noisy, try to increase the strength of the ultrasonic reflection from the target by increasing the target area.

- You may want to show students how to match a distance-time graph one time before asking them to try it on their own. See Part II of the Procedure for guidance.

ANSWERS TO DEVELOPING THE MODEL

If this activity is performed before students have studied motion in one dimension (*Holt Physics*, Chapter 2), students may have difficulty generating these graphs. Helping students think intuitively about how motion can be represented graphically now will help prepare them for more complex analysis later in the course.

1. Distance-time graphs:

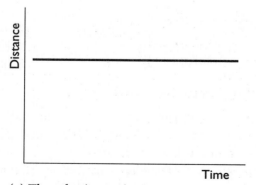

(a) The robotic arm is at rest.

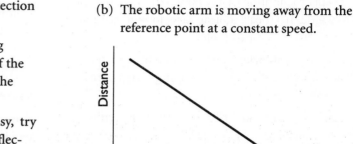

(b) The robotic arm is moving away from the reference point at a constant speed.

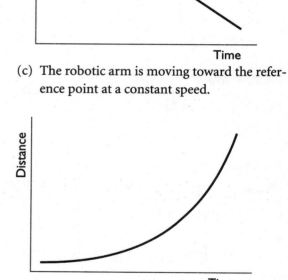

(c) The robotic arm is moving toward the reference point at a constant speed.

(d) The robotic arm is accelerating away from the reference point, starting from rest.

2. Velocity-time graphs:

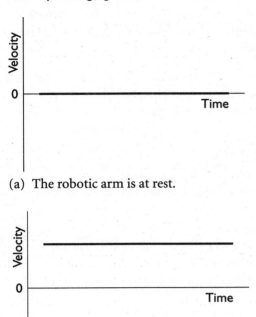

(a) The robotic arm is at rest.

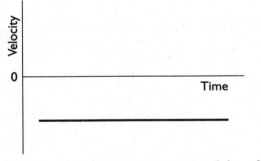

(b) The robotic arm is moving away from the reference point at a constant speed.

(c) The robotic arm is moving toward the reference point at a constant speed.

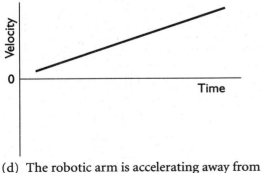

(d) The robotic arm is accelerating away from the reference point, starting from rest.

CHECKPOINTS

Steps 10–12: Students should use the graphs generated by their own calculators in these steps, not the sample graph shown in the instructions. Sketching the graph and writing a description of the motion it represents will prepare them for their attempts to match the graph.

Step 13: Students may find it difficult to match graphs. Encourage students to repeat data collection until they get acceptable results. It helps to place the calculator so that the student can easily see the screen while matching the graph. You can also use a view screen to project the calculator screen image.

Steps 18–19: In general, it is harder for students to match the velocity graphs. One suggestion is for students to use consistent, small footsteps. This approach will allow them more control in achieving the proper speed.

POST-LAB

Cleanup and Disposal
Very little cleanup is required. If the experiment will not be performed again, store equipment securely and remove any tape that is left on the floor or tables. Cardboard may be recycled or saved for use in other activities.

ANSWERS TO ANALYSIS

Part I
1. The graph should slope gradually upward as time increases to the right.

2. You would have to move toward the motion detector at a constant speed.

3. The graphs are not always smooth and even. The appearance of spikes probably indicates that the motion detector is detecting an object other than the intended target, such as a chair or desk, a person walking by, or the wall or ceiling behind the target.

Part II

4. Answers will vary depending on the random graphs generated by the calculator.

5. A positive slope indicates that the robotic arm is moving away from the reference position. A negative slope indicates that the robotic arm is moving toward the reference position.

6. When the slope of a distance-time graph is zero, the robotic arm is at rest.

7. On a distance-time graph, the greater the slope, the greater the speed of the robotic arm.

8. When the slope of the distance-time graph is changing, the robotic arm is accelerating.

Part III

9. Answers will vary depending on the random graphs generated by the calculator.

10. When the slope of a velocity-time graph is zero, the robotic arm is traveling at a constant speed either toward or away from the reference point (unless the velocity is constant at zero, in which case the arm is at rest).

11. When the slope of a velocity-time graph is not zero, the robotic arm is accelerating.

ANSWERS TO CONCLUSIONS

12. Answers will vary, although most students find the distance-time graphs much easier to interpret.

13. Answers will vary. Most students will find the distance-time graphs to be a more useful and effective way to store instructions for the robotic arm.

14. If production increased by 10 percent, the distance-time graphs would have the x-axis compressed by 10 percent. Visually, the graph would appear to be squeezed to the left. Alternatively, students may answer that lines or curves on the graph would have steeper slopes.

15. If production increased by 10 percent, the velocity-time graphs would appear similar. However the graphs would appear to be stretched by 10 percent in a vertical direction. Positive points would be moved upward, and negative points would be moved downward.

EXTENSIONS

1. Make sure students devise tasks involving only one-dimensional (back-and-forth) motion. Otherwise, the motion will be too hard to represent graphically and too complex to capture with the motion detector.

2. Students can extend the reasoning and techniques they have learned in this lab to analyze the behavior of a pendulum qualitatively. Students should be able to recognize that the velocity of a pendulum is not constant and that the velocity is greatest at the bottom of the pendulum's swing.

Technology Lab A

Acceleration

OBJECTIVES

In this lab, students will:

- **Model** the motion of a bungee jumper and use an accelerometer to measure accelerations.
- **Compare** the data from the model with an actual bungee jump.
- **Propose** suggestions to improve the model.

PLANNING

Recommended Time

1 lab period

Materials

[**for each lab group**]

✔ graphing calculator with link cable

✔ CBL system

✔ PHYSICS application loaded in calculator

✔ Vernier low-g accelerometer with CBL adapter cable

✔ table clamp with rod

✔ small plastic bottle with gravel inside

✔ bottle lid with eyebolt, nut, and washer

✔ 0.5 m stretchable cord (bungee cord, a long rubber band or several rubber bands tied together, latex, waistband elastic, surgical tubing, etc.)

✔ meterstick

✔ tape

Classroom Organization

This activity should be performed in groups of two or more.

Materials Preparation

You may want to set up lab stations beforehand. If you have more than one lab class, the equipment can be left standing from one class to the next.

A support stand clamped to the table may be used instead of a table clamp and rod. In either case, the support must be strong and stable, to withstand the forces exerted by the cord during the experiment. Check stands periodically during the activity.

Small plastic bottles filled with gravel work very well for jumpers. Drill a small hole in the lid and attach a small eyebolt. Use a wide nut and a washer inside the lid to keep the bolt from pulling out. This set up allows gravel to be added or subtracted until the weight is appropriate.

Provide students with a variety of cords. Small-diameter surgical tubing will approximate the real bungee jump data. You can also test elastic waistband material, actual bungee cord, and rubber bands. Inspect all cords before use. Discard any cords that show wear or fraying.

For some cord materials, a longer length combined with a greater mass may work better. You may wish to build a launching platform that allows the jumpers to be released from a height that is greater than that possible with a support stand on a tabletop. If you do, make sure the platform is sturdy, secure, and safely accessible.

🔌 internet **connect**

**go.
hrw
.com**

GO TO: go.hrw.com
KEYWORD: HF2 TECHLABS
This lab may also be performed with TI CBL 2 or Vernier LabPro systems. Instructions are available at this Web site.

Techniques to Demonstrate

Demonstrate setting up the apparatus so that the jumper will not hit the floor when it is dropped. Run through the entire procedure once for the class, emphasizing safety concerns at each point.

Use a sample data set or the graph of actual bungee jump data to review acceleration graphs. Students may mistakenly try to analyze the acceleration graph as if it were a record of the jumper's position.

ANSWERS TO DEVELOPING THE MODEL

1. Answers will vary depending on the material used. Students' descriptions should compare the original length to the stretched length of a section of the cord.

2. Students should describe the displacement of the jumper. Descriptions could include references to the height of the jumper when it is released, at the lowest point, and at the highest point during a bounce.

3. Students should explain that they could tell that the cord had lost the tension because the cord was loose or was not completely straight at some point during the bounces.

CHECKPOINTS

Step 1: Students should attach the accelerometer securely. Make sure the accelerometer is attached so that the arrow will point up throughout the entire fall.

Step 2: Make sure students attach the cord securely to the top of the bottle so that the jumper will not come loose. Students should be able to demonstrate that the jumper will not hit the floor when it released from the selected height. Make sure students lay the cord flat or otherwise remove tension from the cord before measuring its length.

Step 7: Have students practice catching the jumper several times before they begin taking data. If students do not catch the jumper during the data collection, help them identify the free-fall values on the graph.

POST-LAB

Cleanup and Disposal

Very little clean up is required. If the experiment is not to be performed again, store equipment securely. You may wish to save rubber bands and other elastic materials for future use. However, it is important to inspect or test these materials before each use, as they will degrade over time.

DATA TABLE WITH SAMPLE DATA

Time (s)	Acceleration (m/s^2)	Direction of motion (up, down, or rest)
0.5	0.0	rest
0.92	−9.6	down
1.16	0.6	down
1.5	18	rest
1.9	0.4	up
2.36	−9.6	rest
2.9	8.9	down
3.26	9.8	up

ANSWERS TO ANALYSIS

1. Sample data is shown in the data table.

2. Refer to the sample graph below. At the first point when the bungee has no slack, the downward acceleration begins to increase from −9.8 m/s^2.

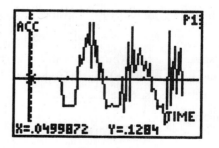

3. Refer to the sample graph above. The moment when the bungee is stretched to its maximum can be found by looking for the point at which the upward (positive) acceleration is the greatest.

4. The maximum acceleration for the sample data was about 18 m/s² upward. This acceleration was reached when the cord was stretched to its maximum. The maximum for the real jump was also about 18 m/s² upward.

5. When the acceleration is at its maximum, the bottle is at the lowest point in the fall. The bottle's velocity is zero at this point.

6. When there is no tension in the cord, the bottle undergoes an acceleration of −9.8 m/s² over a short but continuous period of time. The highest point would occur near the middle of this time period.

ANSWERS TO CONCLUSIONS

7. Answers will vary. The sample data shown are very similar to the real jump data. In both cases, two bounces occurred. The maximum accelerations achieved on the first bounce were similar. The duration of the jump was much shorter on the laboratory model because the cord that was used was much shorter. The laboratory model appears to be useful for analyzing the acceleration rates and the effects of bouncing.

8. **a.** With a heavier jumper, the cord will stretch more. Thus, the time scale may be longer for the jump. It is hard to predict how a heavier jumper would affect the acceleration experienced. In general, small increases in mass will result in a greater force acting on the jumper, but not necessarily a greater acceleration. On the other hand, large increases in weight may result in smaller maximum acceleration because the cord may be stretched beyond its ability to bounce back.

b. With a longer cord, the time during free fall would be extended and there would be a greater time between the greatest negative acceleration and the greatest positive acceleration.

c. If the cord required more force to stretch it, the events of the jump would occur in a smaller period of time. The jumper would fall for a shorter length of time. The acceleration values during periods when the cord is stretched would be greater.

EXTENSIONS

1. The cord used in the real bungee jump was 20 m long. Students should explain whether performing the calculation on their results matched the measured length of their cord.

2. Answers will vary. Students may suggest safety measures such as using different types of cord for jumpers in different weight groups or setting up the equipment so that all jumpers will be able to jump safely.

3. Student answers should cite references and should compare the accelerations experienced by astronauts and those experienced by bungee jumpers. Accelerations of 6 *g* are considered to be the maximum that can be tolerated by astronauts.

4. Answers will vary. Students may say that the accelerometer is a better tool because they can directly analyze the acceleration without performing calculations. Other students may prefer the motion detector because it allows them to correlate the path of the jumper to the shape of the graph more directly.

Technology Lab B

Chapter **2**

Free Fall

Objectives

In this lab, students will:

- **Measure** the acceleration of several falling objects of different masses.

- **Compare** measured free-fall accelerations with an accepted standard value.

- **Determine** whether or not the free-fall acceleration of an object depends directly on the object's mass.

Planning

Recommended Time

1 lab period

You may do this lab in a shorter time, with a more focused result, by performing Part I of the procedure alone. The method in Part I provides a precise and accurate measurement of *g*.

Materials

[for each lab group]

✔ graphing calculator with link cable

✔ CBL system

✔ PHYSICS application loaded in calculator

✔ Vernier photogate with CBL adapter

✔ Vernier picket fence

✔ table clamp and rod

✔ support ring with clamp

✔ additional clamp (for attaching photogate)

✔ 1 m piece of string

✔ 4–5 assorted objects, all about the same size and shape as a baseball: hollow plastic ball, crumpled newspaper, partially inflated balloon, baseball, etc.

✔ cushion, pillow, or padded catch box

✔ strong, clear tape

Classroom Organization

This activity should be performed in groups of two or more.

Materials Preparation

You may want to set up lab stations beforehand. If you have more than one lab class, the equipment can be left standing from one class to the next.

A support stand clamped to the table may be used instead of a table clamp and rod. In either case, the support must be strong and stable, to withstand the force exerted by objects falling onto the support ring.

Photogates should be mounted well below the level of the eyes and heads of students.

Picket fences are available from Vernier Software and other supply companies. You can also make your own picket fences using clear plastic and black electrical tape. However, the precision of your results depends on the precision of the distance between the bars on the fence. The PHYSICS application assumes a leading edge separation of 5.00 cm.

A catch box can be made by lining the bottom of a cardboard box with foam. Catch boxes or cushions should be under the picket fence throughout lab. In Part II, tape may pull free of the picket fence so that it falls to the floor.

⤤ internet connect

GO TO: go.hrw.com
KEYWORD: HF2 TECHLABS
This lab may also be performed with TI CBL2 or Vernier LabPro systems. Instructions are available at this Web site.

Techniques to Demonstrate

Demonstrate dropping the picket fence vertically and without rotation. This is best achieved by holding the picket fence between two fingers at the very center of the top edge. Release the picket fence by pulling your fingers apart.

ANSWERS TO DEVELOPING THE MODEL

1. The other required piece of information is the distance between the leading edges of each of the black bars on the picket fence.

2. If the picket fence goes through the photogate at an angle, the beam will be interrupted at longer time intervals. The resulting velocity and acceleration values will be lower.

3.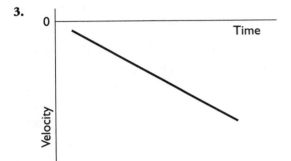

The shape of a velocity-time graph for an object in free fall is a straight line that crosses the y-axis close to the origin. The fact that the line is straight suggests that the acceleration is constant.

4. The units of slope on a velocity-time graph are m/s^2. This slope represents acceleration, as the units suggest.

CHECKPOINTS

Step 8: Each time students generate a velocity-time graph, they must perform a linear curve fit (step 1 of the Analysis) before obtaining more data. This step is performed five times for each object tested by the students.

Steps 10–14: If students are only performing Part I of the Procedure, they may skip these steps and proceed directly to step 2 of the Analysis (step 1 should already have been completed). If using this option, students should do all of the Analysis, but only step 5 of the Conclusions.

Step 12: Cushions or catch boxes should still be used for Part II because the picket fences may fall if strings break or if tape comes loose.

DATA TABLES WITH SAMPLE DATA

Object	Mass (kg)	Slope (trial 1)	Slope (trial 2)	Slope (trial 3)	Slope (trial 4)	Slope (trial 5)
picket fence alone	0.083	9.797	9.791	9.810	9.810	9.819
plastic ball	0.119	8.441	8.339	8.315	8.499	8.510
crumpled paper	0.096	7.050	7.155	6.952	7.205	7.050
baseball-sized balloon	0.097	6.121	6.001	6.258	6.158	6.197
baseball	0.236	9.014	9.254	9.198	9.065	9.300

Object	Average slope	Free-fall acceleration (m/s^2)	Percent uncertainty
picket fence alone	9.802	-9.80 ± 0.01	0.1 %
plastic ball	8.421	-8.4 ± 0.1	1 %
crumpled paper	7.082	-7.1 ± 0.3	4 %
baseball-sized balloon	6.147	-6.1 ± 0.3	5 %
baseball	9.164	-9.2 ± 0.3	3 %

POST-LAB

Cleanup and Disposal

Very little clean up is required. If the experiment is not to be performed again, store equipment securely.

ANSWERS TO ANALYSIS

1. Answers will vary depending on data. Slopes for the sample data are shown in the first sample data table. Note that the PHYSICS application records all the velocity values as positive, and therefore generates positive slopes.

2. Answers will vary depending on data. Averages for the sample data are shown in the second sample data table.

3. Answers will vary depending on data. Make sure students are using half the difference between maximum and minimum values, not the whole difference. The column *Free-fall acceleration* in the second sample data table shows *average slope* (rounded) ± *uncertainty*.

4. Answers will vary depending on data. Percent uncertainties for the sample data are shown in the sample data table.

ANSWERS TO CONCLUSIONS

5. Answers will vary. For the sample data, the accepted value of −9.81 m/s² is on the edge of the range of values for the free-fall acceleration of the picket fence alone.

6. Answers will vary. Objects with high mass and low drag will have free-fall accelerations closest to the free-fall value of the picket fence alone. For the sample data, the baseball was the closest and the balloon was the farthest.

7. Answers may vary, but in most cases, the students should find that not all objects fall at the same rate of acceleration, even accounting for uncertainty. This appears to contradict Galileo's hypothesis.

8. No, he would not be able to demonstrate that all objects fall at a uniform rate with these two objects. The baseball would strike the ground first.

9. Answers will vary. Proposed explanations should include the possibility that air resistance or drag is slowing the rate of acceleration for some objects more than for others.

10. A strong counterexample to Aristotle's hypothesis would show a lighter object falling at a greater rate of acceleration than a heavier object. Students results may or may not show a clear example of this. If they do, then the data contradict Aristotle's hypothesis. If they do not, the hypothesis is neither supported nor refuted.

EXTENSIONS

1. Students may extend their analysis of the relationship between free-fall acceleration and mass by plotting a graph of free-fall acceleration versus mass. If there is a directly proportional relationship, as Aristotle believed, then the data should have a linear relationship on the graph.

2. Answers should show a graph with displacement or distance on the vertical axis and time on the horizontal axis. The curve should be parabolic and should fit the data points fairly closely.

3. Students should find that free-fall acceleration, *g*, varies only slightly with altitude on Earth's surface. At an altitude of 9000 m, about the height of Mt. Everest, *g* is less than 1 percent lower than *g* at sea level. Other factors that may affect *g* include the density of Earth below a given point.

Technology Lab

Chapter 3

Projectile Motion

OBJECTIVES

In this lab, students will:

- **Develop** a model that can be used to simulate a car driving off a cliff on a movie set.
- **Analyze** the relationship between the horizontal velocity and impact point of a projectile, using this model.
- **Evaluate** the assumptions made by the model.
- **Predict** the velocity required for the car to land at a targeted spot on the beach.

PLANNING

Recommended Time

1 lab period

Materials

[for each lab group]

✔ graphing calculator with link cable
✔ CBL system
✔ PHYSICS application loaded in calculator
✔ two Vernier photogates
✔ Vernier dual photogate adapter
✔ two support stands
✔ two right-angle clamps
✔ steel ball (1 to 3 cm diameter)
✔ ramp (angle molding, clay, and wood blocks or additional support stand)
✔ plumb bob
✔ meterstick or metric measuring tape
✔ piece of cardboard or catch box
✔ masking tape
✔ carbon paper

Classroom Organization

This activity should be performed in groups of three or more.

If possible, the equipment should be set up so that the balls travel away from the center of the room; runaway balls on the floor can be a safety hazard.

To help students prepare for the lab, assign Developing the Model questions for completion prior to entering the lab room.

Materials Preparation

You may want to set up lab stations beforehand. If you have more than one lab class, the equipment can be left standing from one class to the next.

The ramps can be made with angle molding, corner molding, metal U-ball molding, etc. Use clay to hold the molding in place on a stack of wood blocks or on a horizontal rod extended from a support stand. To reduce bouncing, you can file the bottom of the ramps to make them lie more level on the tabletops.

Instead of plain pieces of cardboard, you may provide cardboard boxes for catching balls after they strike the floor.

▸ internet connect

GO TO: go.hrw.com
KEYWORD: HF2 TECHLABS
This lab may also be performed with TI CBL 2 or Vernier LabPro systems. Instructions are available at this Web site.

Techniques to Demonstrate

You may want to show students how to release a ball from the ramp without pushing it, so that the velocity of the ball at the bottom of the ramp is due entirely to the ball's motion down the ramp.

ANSWERS TO DEVELOPING THE MODEL

1. Factors that will affect the time the car takes to hit the beach include the height of the cliff and the amount of air resistance. Similar factors affect the ball; however, the car is not as symmetrical as the ball, and air resistance will differ.

2. The relationship between horizontal velocity and horizontal distance can be described as a directly proportional relationship. The greater the horizontal velocity, the greater the distance the ball would travel. The actual distance can be calculated as $v_0 \Delta t$, where Δt is the time it takes the car to fall the height of the cliff. This relationship could be shown as in the following graph, where Δt is the slope of the line.

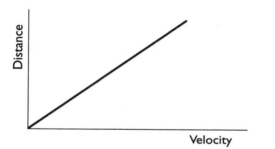

3. If you know the time interval between the two photogates, Δt, and the distance between the two photogates, d, you can calculate the speed of the ball as $d/\Delta t$.

CHECKPOINTS

Step 3: If time allows, more than three positions on the ramp can be used to provide more data points.

Step 4: If the distance between the photogates is not measured accurately, all speed measurements will be systematically too high or too low. This distance measurement should be made carefully and with the photogates parallel to each other. If the photogates are moved during the experiment, this distance measurement will have to be taken again.

Step 5: The measurement of Δy should be made from the top of the table.

POST-LAB

Cleanup and Disposal

Very little cleanup is required. If the experiment is not to be performed again, store equipment securely and remove any tape that is left on the floor or tables. Dispose of carbon paper.

DATA TABLE WITH SAMPLE DATA

Distance between photogates (m), d			0.01			
Table height (m), Δy			0.722			
Launch point	Trial	Time (s), Δt	Horizontal velocity (m/s), v	Horizontal distance (m), Δx	Average velocity (m/s), v	Average distance (m), Δx
Low	1	0.00921	1.085	0.418	1.079	0.422
	2	0.00935	1.069	0.422		
	3	0.00912	1.083	0.425		
Medium	1	0.01105	1.302	0.502	1.300	0.502
	2	0.01122	1.283	0.506		
	3	0.01108	1.315	0.498		
High	1	0.01326	1.562	0.602	1.560	0.605
	2	0.01346	1.539	0.608		
	3	0.01329	1.578	0.598		

ANSWERS TO ANALYSIS

1. See sample data table.

2. See sample data table. The highest point on the ramp should have produced the greatest average velocity.

3. The graph below is based on the sample data. Student graphs should suggest a linear relationship between horizontal velocity and horizontal distance.

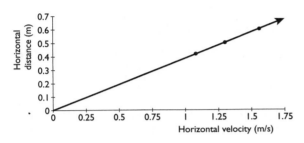

4. The slope for the sample data is 0.387 s.

5. The theoretical slope is 0.379 s.

6. For the sample data, the slope correction factor is 1.02.

7. The units of s_{obs} and s_{th} are seconds. The value of s represents the number of seconds the ball or the car takes to hit the ground.

8. The primary additional factor that affects the flight of the ball is air resistance. Errors in measurements could also account for the difference between theoretical and observed values.

ANSWERS TO CONCLUSIONS

9. The theoretical slope for the car is 4.78 s.

10. Final answers will vary depending on the calculated slope correction factor. The predicted slope for the crash scene based on the sample data will be 4.87 s.

11. Final answers will vary depending on the calculated slope correction factor. The sample data leads to the following recommended speeds: to land 38.0 m from the base of the cliff, the car would have to be traveling 7.80 m/s (17.5 mi/h); to land 42.0 m from the base of the cliff, the car would have to be traveling 8.62 m/s (19.3 mi/h).

12. The experimental setup is similar to the actual scenario in that a rolling object with a horizontal velocity is launched from a surface. Differences include the shape of the falling object, the manner in which speed is measured, and the presence of wind.

EXTENSIONS

1. Students' answers will vary depending on their data. According to the sample data, the film editor could include 3.13 s of footage of the car traveling toward the edge of the cliff.

2. Students' answers will vary depending on their data and the result calculated for the car's speed. The following results are for an initial velocity that would cause the car to land 38.0 m from the base of the cliff, assuming the base of the cliff is the origin. Students' graphs should be parabolic.

Time, Δt (s)	x-coordinate (m)	y-coordinate (m)
0.00	0.00	112.0
0.25	1.9	111.7
0.50	3.9	110.8
0.75	5.9	109.2
1.00	7.8	107.1
1.25	9.8	104.3
1.50	11.7	101.0
1.75	13.6	97.0
2.00	15.6	92.4
2.25	17.6	87.2
2.50	19.5	81.3
2.75	21.4	75.0
3.00	23.4	67.9
3.25	25.4	60.2
3.50	27.3	51.9
3.75	29.2	43.0
4.00	31.2	33.5
4.25	33.2	23.4
4.50	35.1	12.7
4.75	37.0	1.3

3. Students' suggestions for an experimental design will vary. A curved tube can be used to model the ski jump.

Technology Lab A

Static and Kinetic Friction

Objectives

In this lab, students will:

- **Develop** a method for determining the coefficients of static and kinetic friction for shoes on roofing material.
- **Measure** the force of static friction using a force sensor.
- **Determine** if forces of friction or coefficients of friction depend on weight.
- **Recommend** shoes that are suitable for people working on roofs.

PLANNING

Recommended Time

1 lab period

Materials

[for each lab group]

✔ graphing calculator with link cable
✔ CBL system
✔ PHYSICS application loaded in calculator
✔ Vernier force sensor
✔ two shoes (not from the same pair)
✔ composition roof shingles (non-asbestos)
✔ balance
✔ 500 g and 1 kg masses
✔ string
✔ strong tape
✔ graph paper

[for entire class]

✔ balance

Classroom Organization

This activity should be performed in groups of two or more.

Materials Preparation

You may want to set up lab stations beforehand. If you have more than one lab class, the equipment can be left standing from one class to the next.

Strong tape, such as strapping tape or duct tape, should be used. Masking tape may not be strong enough to hold strings to the shoes.

Provide at least two different shoes for each lab group. It is best if the shoes have very different frictional qualities. If possible, provide one or two groups with shoes with high-friction "sticky" rubber soles, such as some new sport sandals, climbing shoes, and hiking boots.

A similar lab may be performed using wooden blocks directly on clean tabletops. In that case, the data will closely match the theoretical models of static and kinetic friction. However, the use of shoes and shingles creates a more "real-world" scenario and reveals the often complex nature of friction. Expect a wide range of coefficients of friction.

GO TO: go.hrw.com
KEYWORD: HF2 TECHLABS
This lab may also be performed with TI CBL 2 or Vernier LabPro systems. Instructions are available at this Web site.

Techniques to Demonstrate

Coach students to increase the force applied to the shoe very slowly and evenly. Their tendency will probably be to increase the applied force too rapidly. The students are also instructed to practice this motion in the Developing the Model exercises.

ANSWERS TO DEVELOPING THE MODEL

1. A graph of the force exerted should show the force first increasing to a peak, then dropping to a constant (horizontal line). A sample sketch is shown below.

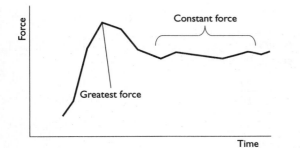

2. See the sketch above. The point marking the greatest force on the graph corresponds to the moment just before the shoe started to move.

3. See the sketch above. During this time, the shoe was moving with a constant velocity.

4. The graph would appear similar except that the scale would be different. Both the force required to move the shoe and the force required to keep the shoe moving would be much greater.

5. For a shoe on a flat floor, the magnitude of the normal force is equal to the magnitude of the force of gravity. For a roofer on an inclined roof, the magnitude of the normal force is equal to the magnitude of a vector component of the force of gravity.

6. The force causing the shoe to move in the exercise was a hand pulling on the string. For a roofer, the force causing the roofer's foot to slip is a vector component of the force of gravity.

7. If someone were to stand in the shoe, the normal force would increase, and the force of friction would also increase. However, for an ideal surface, both forces would increase proportionally, and the coefficients would remain the same.

CHECKPOINTS

Step 1: Students should make two sets of data tables like the ones shown in the Data Tables section, for a total of four tables (two for each shoe).

Step 2: Make sure the tape holds the string securely to each shoe.

Step 7: Students may increase the time between samples to increase the total time in which data are collected.

Step 9: Make sure the shingles are not sliding as the students pull the shoes. One person in each lab group may have to hold the shingle while another person pulls the shoes. A third person could operate the CBL and calculator.

Step 15: If students have additional time, they may take data on additional shoes or on other objects in the laboratory.

POST-LAB

Cleanup and Disposal

Very little clean up is required. If the experiment is not to be performed again, remove the tape and string from the shoes. Store CBL units, sensors, and other equipment securely.

DATA TABLES WITH SAMPLE DATA

1. Sample data are shown below for one shoe only. Students should have data for two shoes (a total of four data tables).

Shoe: ____SHOE I____
Mass of shoe: ____0.457____ **kg**

Total mass (kg)	Normal force (N)	Peak static friction (N)			Average peak static friction (N)
		Trial I	Trial 2	Trial 3	
0.457	4.48	1.59	2.00	1.28	1.60
0.957	9.38	3.33	3.78	3.28	3.46
1.457	14.29	5.00	5.67	4.92	5.20
				μ_s (slope)	0.35

Shoe: ____SHOE I____
Mass of shoe: ____0.457____ **kg**

Total mass (kg)	Normal force (N)	Kinetic friction (N)			Average kinetic friction (N)
		Trial I	Trial 2	Trial 3	
0.457	4.48	0.842	0.795	0.768	0.801
0.957	9.38	1.762	1.720	1.692	1.725
1.457	14.29	2.681	2.751	2.388	2.607
				μ_k (slope)	0.18

ANSWERS TO ANALYSIS

1. Graphs will vary. Two sample graphs are shown below.

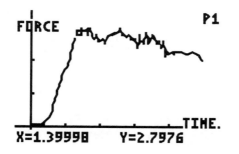

Force-time graph for an empty hiking shoe

Force-time graph for a running shoe with 500 g added

2. Answers may vary. For most shoes, including those used for the sample data, the static friction is greater than the kinetic friction. However, for some very sticky soles, the static and kinetic friction appear to be the same or nearly the same.

3. Answers may vary. Based on the sample data, the coefficient of static friction is greater than the coefficient of kinetic friction. The coefficient of friction is the ratio of the force of friction and the normal force. Because the normal force is constant for each graph, the coefficient of friction will vary the same as the force of friction.

4. Normal forces for the sample data are shown in the sample data tables.

5. Graphs should be roughly linear. A sample graph is shown below.

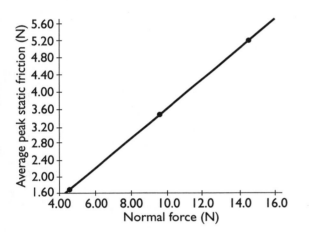

6. For the sample data, $\mu_s = 0.358$.

7. Graphs should be roughly linear. A sample graph is shown below.

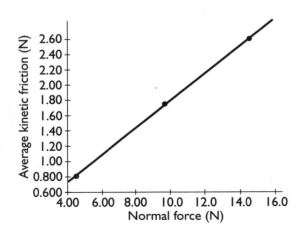

8. For the sample data, $\mu_k = 0.184$.

ANSWERS TO CONCLUSIONS

9. Data should show that the maximum force of static friction and the force of kinetic friction both increase with downward force acting on the shoe (in this case, the combined weight of the shoe and its contents). This is in accordance with the equations for the force of friction, $F_{s,max} = \mu_s F_n$ and $F_k = \mu_k F_n$. As the total weight increases, the normal force on the shoe also increases.

10. Data should show that the coefficients of static and kinetic friction do not depend on the downward force acting on the shoe. Although the forces of friction increase with increasing weight, the normal force also increases proportionally. Therefore, the coefficients of friction remain constant.

11. Given the same kind of shoes, all employees will have an equal risk of slipping on a horizontal surface because the coefficient of friction is independent of the weight acting on a shoe. However, on an inclined surface, a heavier employee will actually have a higher risk of slipping because the force along the incline, which has F_g as a component, will be disproportionately greater than the frictional force for a greater weight on the shoe.

12. Answers will vary. In general, students should recommend the shoe with the highest coefficient of friction, especially the coefficient of static friction. The shoe with the highest coefficient of static friction minimizes the risk of slipping in the first place.

13. Answers will vary. Shoes with high coefficients of friction tend to have complex tread patterns and may use new, high-tech rubber or composite materials. Students should ultimately recommend the shoe with the highest coefficient of friction, especially the coefficient of static friction. Students may also consider other factors, such as stability and comfort.

EXTENSIONS

1. Shoes with a large μ_s and a low μ_k would resist slipping to begin with. However, once you started to slip, you would find it difficult to stop sliding. Shoes with μ_s and μ_k that were nearly the same would be much safer.

2. Predictions will vary, although evidence like that in Demonstration 8 (*Holt Physics ATE*, p. 144) suggests that the force of friction is not closely correlated with the surface area of contact. Experiments that test the hypothesis would test the force required to move shoes with a variety of sole surface areas.

3. Answers will vary depending on the coefficient of friction (determined by the type of shoe and the type of material composing the shingle). The coefficient of friction between the shoe and the shingle should always be the same whether the shingle is inclined or horizontal.

4. Answers will vary. Many shoe manufacturers have developed high-friction soles and continue to research new materials.

Technology Lab B

Air Resistance

OBJECTIVES

In this lab, students will:

- **Develop** a physical model that simulates an object falling with a parachute.

- **Analyze** the relationship between mass and terminal speed of a falling object using data from the model.

- **Evaluate** two mathematical models relating air resistance to terminal speed using data generated in the simulation.

- **Calculate** an air resistance factor for the model.

- **Predict** the required air resistance factor for the parachute and the maximum allowable load for the drop.

PLANNING

Recommended Time

1 lab period

Materials

[for each lab group]

✔ graphing calculator with link cable

✔ CBL system

✔ PHYSICS application loaded in calculator

✔ Vernier motion detector

✔ 5 basket-style coffee filters

✔ balance

✔ graph paper

Classroom Organization

This activity should be performed in groups of two or more (at least three if motion detectors are to be held overhead).

Materials Preparation

You may want to set up lab stations beforehand. If you have more than one lab class, the equipment can be left standing from one class to the next.

The motion detector can be mounted on a high wall bracket, on a ceiling bracket, or using a clamp and rod on a high shelf. The detector should be at least 2 m high. Do not mount at head height.

If an overhead mount is not available, one student may stand on a sturdy chair (without wheels) and hold the motion detector while another drops the filter. The detector must be held as still as possible while data is being collected. Remind students to use caution when standing on chairs and to be careful not to trip over cords.

Another option is to place motion detectors on the floor and drop the filters directly on top of them. The setup is easier in that case, but filters may not reach terminal velocity, and the graphs may be harder for students to interpret because detector readings are unreliable when the filter is closer than 0.4 m.

A CBR (Calculator-Based Ranger) may be used instead of a Vernier motion detector and CBL.

Coffee filters should be flat on the bottom to maximize air resistance. Large commercial-sized basket filters, often used in restaurants and cafeterias, work best.

Instead of coffee filters, you may provide toy parachutes or have students make small parachutes out of handkerchiefs, representing the load with a stopper suspended from the corners of the handkerchief. This works best when the motion detector is positioned above the falling parachute. Do not use lab masses for cargo.

> ⚡ internet **connect**
>
> **GO TO:** go.hrw.com
> **KEYWORD:** HF2 TECHLABS
> This lab may also be performed with TI CBL 2 or Vernier LabPro systems. Instructions are available at this Web site.

Techniques to Demonstrate

You may want to show students how to drop a coffee filter so that it falls straight down and doesn't rotate. A good technique is to hold the filter loosely on either side with the palms of your hands turned vertically. Pull both hands away simultaneously .

If students will need to stand on chairs to hold motion detectors or to drop the filters from sufficient height, demonstrate how to securely position a chair and step up on to it. Caution students to keep their weight over the center of the chair at all times.

ANSWERS TO DEVELOPING THE MODEL

1. The magnitude of the downward force on a falling object is simply the weight, mg, where m is the mass of the object, and g is the acceleration due to gravity. $F_g = mg$.

2. When an object reaches terminal speed, its acceleration is equal to zero. Therefore, the net force on the object must also be equal to zero.

3.

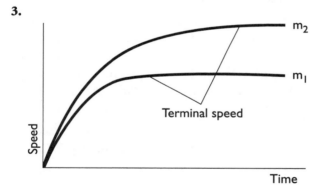

A heavier load has a greater terminal speed than a lighter load.

4. $F_R = -F_g$
$-kv_T = -mg$
$$v_T = \frac{mg}{k}$$

5. $F_R = -F_g$
$-k(v_T)^2 = -mg$
$$v_T = \sqrt{\frac{mg}{k}}$$

CHECKPOINTS

Step 1 The experiment may be performed with the motion detector held overhead by one of the students. In that case, the motion detector must be held as still as possible. The experiment may also be performed with the motion detector on the ground. In that case, students should stand on a chair to drop the filters from as great a height as possible so that terminal speed will be reached before the filter is within 0.5 m of the detector. Make sure students have chairs and other devices positioned safely and that there are no obstructions or protrusions near the level of students' heads.

POST-LAB

Cleanup and Disposal

Very little clean up is required. If the experiment is not to be performed again, store equipment securely.

DATA TABLE WITH SAMPLE DATA

Number of filters	Total mass, m (kg)	Terminal velocity, v_T (m/s)	v_T^2 (m²/s²)	Drag coefficient (kg/m)
1	.0036	1.39	1.93	0.018
2	.0071	1.94	3.76	0.018
3	.0110	2.25	5.06	0.023
4	.0147	2.59	6.71	0.021
5	.0180	2.77	7.67	0.023
			Average =	0.021

ANSWERS TO ANALYSIS

1.

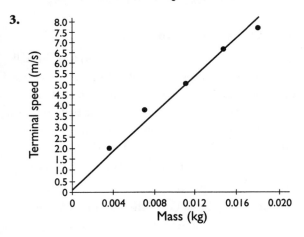

The sample data do not seem to fit a linear model. The origin is not in line with the rest of the data points.

2. Squared terminal speed values for the sample data are shown in the sample data table.

3.

For the sample data, the fit is much better when the terminal speed is squared. The points lie nearly on a line with each other and with the origin.

4. Most students should find that find that the quadratic model (b) fits their data better.

5. See data table for values of k based on the sample data, calculated using the equation $k = \dfrac{mg}{v_T^2}$.

ANSWERS TO CONCLUSIONS

6. The air resistance factor for the parachute would have to be at least 871 kg/m.

7. With the current parachutes, the maximum load would be limited to about 36 kg. Objects weighing 55 kg could not be delivered safely using the current parachutes.

8. The terminal speed reached by the supply box was 49.5 m/s.

9. The coffee filter and the parachute are roughly the same shape. However, the coffee filter was dropped with the convex side facing downward while a parachute falls with the convex side facing upward. This difference would probably affect the air resistance factor but would probably not significantly affect the linear relationship between mass and velocity squared.

EXTENSIONS

1. Check student designs before students proceed. Students should find that a small parachute follows roughly the same mathematical model as the coffee filters.

2. Students must set $F_R = -F_g$ and solve for C_D. Students will also need to measure the diameter of a coffee filter and calculate a cross-sectional area. The cross-sectional area of the filters used for the sample data was 0.022 m^2. Based on the sample data, the drag coefficient, C_D, for a coffee filter is 1.4.

3. Students must set $F_R = -F_g$ and solve for A. Drag coefficients for parachutes can sometimes be found in books and magazines about rocketry. Students may also search the Internet for drag coefficient data.

4. Based on the sample data, a 65 kg human would need to use a coffee filter with a cross-sectional area of 4 m^2 to reach terminal speed at 3 m/s.

Technology Lab

Chapter 5

Loss of Mechanical Energy

OBJECTIVES

In this lab, students will:

- **Measure** the change in the kinetic and potential energy as a ball moves up and down in free fall.
- **Graph** potential energy, kinetic energy, and total energy.
- **Analyze** the graph to determine how much kinetic energy is lost.
- **Reach conclusions** regarding the amount of energy possessed by the volleyball as it fell in the neighbor's driveway.

PLANNING

Recommended Time

1 lab period

Materials

[**for each lab group**]

- ✔ graphing calculator with link cable
- ✔ CBL system
- ✔ PHYSICS application loaded in calculator
- ✔ Vernier motion detector
- ✔ wire basket
- ✔ volleyball or other ball

Materials Preparation

You may want to set up lab stations beforehand. If you have more than one lab class, the equipment can be left standing from one class to the next.

The TI CBR (Calculator-Based Ranger) can be used to replace both the motion detector and the CBL system. Connect the CBR directly to the calculator using a black link cable, and use the PHYSICS application as directed.

internet connect

GO TO: go.hrw.com
KEYWORD: HF2 TECHLABS
This lab may also be performed with TI CBL 2 or Vernier Lab Pro systems. Instructions are available at this Web site.

Classroom Organization

This activity should be performed in groups of three or more. Because this activity involves throwing balls into the air, you may wish to perform the activity outside or in a gym. If you do so, you will need to use batteries instead of AC to power the CBLs.

Techniques to Demonstrate

You may want to show students how to toss the ball for best results. A higher toss results in a greater loss of energy; however, if students throw the ball too high, horizontal motion may carry the ball out of the path of the detector or even out of range. You may not get useful data if your hands are between the ball and the detector for more than about one second. When demonstrating:

- Place a wire basket over the motion detector and make sure all cables are out of the way.
- Hold the ball in both hands, with one hand on either side of the ball.
- Keep the ball directly above the motion detector throughout its motion.
- Have a student ready to catch the ball on its way down, 0.5 m above the motion detector. Once the ball is caught, the student should hold the ball still until the data collection stops.
- If the motion detector does not collect data throughout the toss, you may need to increase the number of intervals so that it will collect data for a longer time.
- Experiment by tossing to different heights in order to find the height that gives the best readings.

ANSWERS TO DEVELOPING THE MODEL

1. As a ball is thrown upward, the potential energy increases linearly with its distance from the reference point. This is shown mathematically with the equation $PE = mgh$. When the ball reaches its highest point, potential energy is at a maximum. As the ball falls, the potential energy decreases as it is converted into kinetic energy. To calculate potential energy, you must know both the mass and the height of the ball (and the free-fall acceleration, g.)

2. When a ball is tossed upward, the kinetic energy of ball is greatest at the moment it leaves the hand. As is rises, it slows down, and the kinetic energy is at a minimum when the ball is at its highest point. If the ball has no horizontal component to its velocity, the kinetic energy is zero at that point. As the ball falls, the kinetic energy increases. To calculate kinetic energy, you must know both the mass and the velocity of the ball. This is shown mathematically in the equation $KE = \frac{1}{2}mv^2$.

3. A graph of velocity versus time should show a line sloping downward. The line should cross the x-axis ($v = 0$) at the time corresponding to the ball being at its highest point. Assuming no air resistance, the line would be straight because an object in free fall is under a constant acceleration. If air resistance is taken into account, the graph will not show a straight line because air resistance will slow the rate of acceleration depending on the velocity.

4. The graph of vertical distance versus time should show a parabola that opens downward. If you assume there is no air resistance, the graph will be symmetrical. If air resistance is taken into account, the graph will not be symmetrical.

5. Answers will vary. The biggest loss of kinetic energy will most likely be due to transfer of energy to air molecules (because of air resistance). These could have a measurable effect on the model depending on the type of ball used and the sensitivity of the motion detector.

CHECKPOINTS

Step 6: Check distance-time graphs to make sure they show a smooth region of rising and falling. If the graphs are very messy, the students' hands may be getting in the way while collecting data. When using the motion detector, it is important to realize that the ultrasound is emitted in a cone about 30° wide. Anything within the cone of ultrasound can cause a reflection and possibly an accidental measurement. Have students repeat the trial if the graphs are not clear.

POST-LAB

Cleanup and Disposal

Very little clean up is required. If the experiment is not to be performed again, store equipment securely.

DATA TABLE WITH SAMPLE DATA

Mass of ball: ____0.311____ kg

Position	Time (s)	Height (m)	Velocity (m/s)	PE (J)	KE (J)	ME (J)
Just after release	1.15	0.907	2.32	2.75	0.837	3.60
Between release and top	1.25	1.07	1.40	3.26	0.305	3.56
Top of path	1.40	1.13	−0.066	3.44	0.0006	3.44
Between top and catch	1.60	0.885	−1.91	2.69	0.567	3.25
Just before catch	1.7	0.621	−2.92	1.89	1.32	3.21

ANSWERS TO ANALYSIS

1. The upper and lower bounds will vary depending on data. For the sample data, the desired time range was from 1.15 to 1.70 s, as shown in the sample data table.

2. An example of a truncated velocity-time graph for a ball toss is shown below. Student graphs will vary depending on data. Curves should be roughly linear, but should also show a slight curvature as the speed of the ball decreases due to air resistance. Velocity values for sample data are shown in the sample data table.

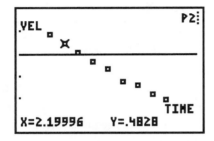

3. An example of a distance-time graph for a ball toss is shown below. Student graphs will vary depending on data. The curves on the graphs should be roughly parabolic.

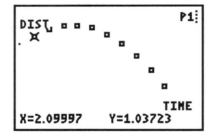

4. Answers will vary depending on data. Total mechanical energy should always be the sum of the potential and kinetic energy for each point. Energy values for sample data are shown in the sample data table.

5. Kinetic energy values will be stored in L_2, potential energy values will be stored in L_3, and total energy values will be stored in L_6.

6. An example of a graph showing energy values for a ball toss is shown below. Student graphs will vary depending on data. The curve for kinetic energy should be bowl-shaped, with the bottom point touching the zero line near the middle of the graph. The curve for potential energy should be shaped like an inverted bowl, mirroring the kinetic energy curve and lying above it. The curve for total mechanical energy should be roughly linear, sloping slightly downward.

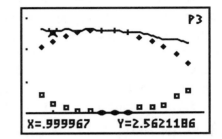

ANSWERS TO CONCLUSIONS

7. Answers will vary depending on the graphs originally sketched in Developing the Model and on student data. The two sets of graphs may differ significantly if air resistance was not taken into account in the initial sketches.

8. In general, students should find that the total mechanical energy decreases over time. This is supported by the slightly negative slope of the line representing total mechanical energy.

9. Answers will vary depending on data. In general, the total energy lost will be around 10 to 20 percent. For the sample data, 10.8 percent of the total mechanical energy was lost during the flight of the ball.

10. If we assume that the volleyball lost 15 percent of its total mechanical energy during its flight, the ball would have had 16.7 joules of kinetic energy when it hit the car.

11. Answers will vary depending on data, although most students should find that the ball would not have dented the roof. Using the 10.8 percent energy loss from the sample data, the approximate kinetic energy of the ball at the time of impact would be 17.5 J.

EXTENSIONS

1. Depending upon the horizontal component of velocity, the ball may have had considerably more kinetic energy. This might suggest that the

ball was more likely to cause a dent, but there may also be other factors to consider. For example, in a case of significant horizontal velocity, the volleyball would hit the roof of the car at an oblique angle and might therefore be less likely to cause a dent.

2. A beach ball would lose more kinetic energy than a volleyball following the same trajectory. The effect of air resistance would be greater on a lighter ball than on a heavy one.

3. If you had entered the wrong mass into the calculator, the relative values for energy would have remained the same. Conclusions would not be affected.

4. This extension exercise can also be done as a supplement to Chapter 6 after students have read the section on elastic and inelastic collisions.

Technology Lab

Impulse and Momentum

OBJECTIVES

In this lab, students will:

- **Develop a model** that demonstrates the impulse-momentum theorem.
- **Measure** momentum change and impulse using the model under various conditions.
- **Relate the model** and the impulse-momentum theorem to military pilots and the stopping distances of jets.

PLANNING

Recommended Time

1 lab period

Materials

[for each lab group]

- ✔ graphing calculator with link cable
- ✔ CBL system
- ✔ PHYSICS application loaded in calculator
- ✔ Vernier motion detector
- ✔ Vernier force sensor and CBL adapter
- ✔ support stand and clamp
- ✔ dynamics cart
- ✔ elastic cord
- ✔ string
- ✔ 500 g mass

Classroom Organization

This activity should be performed in groups of two or more.

Materials Preparation

You may want to set up lab stations beforehand. If you have more than one lab class, the equipment can be left standing from one class to the next.

Many dynamics cart sets come with elastic cord. A good alternative elastic cord for this experiment is the cord used on name badges or the rubber band from a paddle-ball toy.

↗ internet connect

go. hrw .com

GO TO: go.hrw.com
KEYWORD: HF2 TECHLABS
This lab may also be performed with TI CBL 2 or Vernier LabPro systems. Instructions are available at this Web site.

Techniques to Demonstrate

You may want to show students how to release the cart for best results. Push the cart so it rolls toward the motion detector, stretches the elastic band, and gently bounces back to your hand. Make sure students understand that this is a gentle motion, unlike an actual aircraft landing. The force sensor must not shift, and the cart must move in a straight line. Arrange the cord and string so that when they are slack they do not interfere with the cart motion. You may need to guide the string by hand, but be sure that you do not apply any additional force to the cart or force sensor. Make sure your hands stay out of the space between the cart and the motion detector.

ANSWERS TO DEVELOPING THE MODEL

1. Because the final speed, v_f, is 0 m/s, the first term on the right side will drop out, and the equation can be written:

$$F\Delta t = \Delta p = -mv_i$$

2.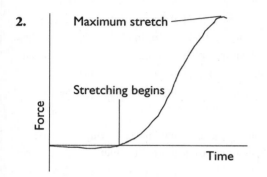

3. The forces acting to slow down a jet include the elastic force exerted by the spring line, the frictional force applied by the brakes, and the force of friction between the wheels and the runway. For the cart, the primary force will be the elastic cord. The model with the cart is a much simpler system than the actual scenario.

4. Engineers would have to take into account the mass of the jet, the length of the runway, and the force exerted by the elastic cord. These factors could be tested by varying the amount of mass added to the cart, measuring the displacement of the cart, and varying the kind or number of cords used in the experiment.

CHECKPOINTS

Step 2 Use holes on the cart end to attach the cord. If the cord is long enough and flexible enough, the string may be omitted.

Step 3 Inspect student setups to make sure the force sensor, the cart, and the motion detector are aligned properly and that all attachments are secure. Make sure the motion detector is far enough away that the cart will not reach within 0.4 m of the detector. Periodically check the connections and cords for wear.

Step 7 The force sensor will give adequate results without calibration, but if it is calibrated the results will be improved.

Step 13 In the sample data below, the first set was collected using a thin rubber band. The second set of data was collected using two parallel rubber bands.

DATA TABLES WITH SAMPLE DATA

Mass of cart: _____0.500_____ kg

Cord	Trial	Initial velocity, v_i (m/s)	Final velocity, v_f (m/s)	Change in velocity, Δv (m/s)	Stopping distance, d (m)	Average force, F (N)	Duration of impulse, Δt (s)	Impulse (N•s)
elastic 1	1	−0.759	0	0.759	0.62	1.058	0.66	0.698
	2	−0.925	0	0.925	0.72	1.197	0.70	0.837
elastic 2	1	−0.790	0	0.790	0.41	1.62	0.44	0.713
	2	−0.828	0	0.828	0.48	2.01	0.38	0.764

Cord	Trial	Impulse, $F\Delta t$ (N•s)	Change in momentum, Δp (kg•m/s or N•s)	Percent difference between $F\Delta t$ and Δp
elastic 1	1	0.698	0.704	0.9%
	2	0.837	0.859	3%
elastic 2	1	0.713	0.688	4%
	2	0.764	0.739	4%

POST-LAB

Cleanup and Disposal

Very little clean up is required. If the experiment is not to be performed again, store equipment securely.

ANSWERS TO ANALYSIS

1. Answers will vary depending on data. Initial velocities for the sample data are shown in the first sample data table. Note that all velocity values are given as negative when the cart is moving toward the motion detector.

2. Answers will vary depending on data. Note that the force values are positive, which is consistent with the direction determined by the motion detector (positive away from the detector).

3. Answers will vary depending on data. Stopping distances for sample data are shown in the first data table. Students may have trouble understanding how to determine stopping distance from the data. Points on the distance–time graphs will be given as distances from the motion detector. To find the stopping distance, students will have to identify the point where the cart starts its motion and the point where its speed is 0 m/s, and then take the *difference* between the two distance values on the graph.

4. Answers will vary depending on data. Δv should always be equal to $-(v_i)$. Values of Δv for sample data are given in the first data table.

5. Answers will vary depending on data. Impulse values for sample data are shown in the second data table. Values should have only three significant figures.

6. Answers will vary depending on data. Values for the change in momentum for sample data are shown in the second data table. Values should have only three significant figures.

ANSWERS TO CONCLUSIONS

7. Answers will vary depending on data. Percent differences for sample data are shown in the second data table. For the sample data, all trials showed less than 5 percent discrepancy between impulse and change in momentum.

8. The model is consistent with the impulse-momentum theorem. The data show that values for impulse and change in momentum, measured independently, are in close agreement (less than 10 percent discrepancy). If students find a larger discrepancy, they may conclude that the experiment does not provide supporting evidence for the theorem.

9. The stopping distance appears, roughly, to be inversely proportional to the average force. More data would be required to firmly establish this relationship.

10. A jet with mass and velocity has momentum. In order to stop the jet, the momentum must be reduced to zero. By using a spring line to apply a force to the jet for a period of time, an impulse is generated that stops the jet. For a jet that is landing on a short runway, the momentum must be reduced to zero very quickly. This would require a greater force acting over a shorter period of time.

11. Answers will vary. Key points would include explaining the momentum of a jet, the impulse required to bring the jet to a stop, and the relationship between average force, duration of force, and stopping distance.

EXTENSIONS

1. These two extensions may be done together, so that students perform research and then make a presentation combining their research and their conclusions following the experiment.

2. A poster presentation provides a concrete way to assess students' understanding of impulse and momentum. This assignment also helps develop design and communication skills.

Technology Lab

Centripetal Acceleration

OBJECTIVES

In this lab, students will:

- **Develop** a model to measure acceleration using a turntable.

- **Determine the relationship** between centripetal acceleration, radius, and angular velocity using the model.

- **Calculate** the radius and angular velocity to be used in new amusement park rides based on their data.

PLANNING

Recommended Time

1 lab period

Materials

[for each lab group]

✔ graphing calculator with link cable

✔ CBL system

✔ PHYSICS application loaded in calculator

✔ Vernier low-*g* accelerometer

✔ Vernier adapter cable

✔ turntable with three speed settings

✔ meterstick

✔ masking tape

✔ stopwatch or watch with second hand

✔ 20 g slotted mass

[for entire class]

✔ small level

Classroom Organization

This activity should be performed in groups of two or more.

Materials Preparation

You may want to set up lab stations beforehand. If you have more than one lab class, the equipment can be left standing from one class to the next.

Safety is an important issue in this lab. Make sure the accelerometer, the CBL, and the cable joining them are all taped down securely. Also tape down or remove the tone arm of the turntable so it does not interfere with the motion of the turntable and the objects on it.

The turntables should have three different speeds in order to produce meaningful results in the Analysis. Turntables with three speeds can still sometimes be found in thrift stores. You may also check your school's AV center or ask other teachers if they have an older turntable with three speeds.

A cardboard circle can be cut out and placed on top of the turntable platter. You can then tape the accelerometer, CBL, and cables onto the cardboard without gumming up the turntable.

---

internet connect

GO TO: go.hrw.com
KEYWORD: HF2 TECHLABS
This lab may also be performed with TI CBL 2 or Vernier Lab Pro systems. Instructions are available at this Web site.

Techniques to Demonstrate

Explain that the CBL will be disconnected from the calculator when collecting data using the turntable. Point out to students the importance of paying attention to the direction of the arrow on the accelerometer. The accelerometer arrow should point directly toward the center of the turntable.

ANSWERS TO DEVELOPING THE MODEL

1. The speed of the mass is constant, but the velocity is changing because the mass is changing its direction of motion. The acceleration is also changing because the acceleration is changing its direction. The direction of the acceleration is always toward the center of the turntable.

2. The mass undergoes a greater acceleration when the turntable is rotating at 45 rpm than when it is rotating at 33⅓ rpm. Proposed mathematical relationships may vary, but should suggest that centripetal acceleration increases as angular velocity increases. Two possibilities are that centripetal acceleration is directly proportional to angular velocity or that centripetal acceleration is directly proportional to the square of angular velocity (the second one is true).

3. Speeding up while moving in a positive direction would result in a positive acceleration. Slowing down while moving in a positive direction corresponds to a negative acceleration.

CHECKPOINTS

Step 8 Make sure the accelerometer, the CBL, and the cable joining them are all taped down securely. Also tape down or remove the needle arm so it does not interfere with the motion of the objects on the turntable.

Steps 8, 18, and 24: The arrow of the accelerometer must always point directly inward toward the turntable spindle; if it does not, the acceleration values will be systematically low.

Steps 13 and 21: Remind students to disconnect the CBL from the calculator before starting the rotation of the turntable.

POST-LAB

Cleanup and Disposal

Very little clean-up is required. If the experiment is not to be performed again, remove any tape that is on the turntable and store all equipment securely. Recycle cardboard or leave the cardboard circles on the turntables for protection and future use.

DATA TABLES WITH SAMPLE DATA

Part I

Acceleration relative to arrow on the accelerometer (positive or negative)	positive

Part II

Radius (m)		0.115	
Angular speed (rpm)	Angular speed (rad/s)	(Angular speed)^2 (rad/s)^2	Centripetal acceleration (m/s^2)
33⅓	3.49	12.2	1.40
45	4.71	22.2	2.56
78	8.16	66.6	7.76

Part III

Radius (m)	Centripetal acceleration (m/s^2)
0.115	7.76
0.085	5.84
0.044	3.10

ANSWERS TO ANALYSIS

1. See the sample data table for Part II for values of angular speed and angular speed squared for the three different turntable speed settings.

2. Graphs will vary depending on data. A graph based on the sample data is shown below. The graph suggests that there is a linear relationship between centripetal acceleration and angular speed squared.

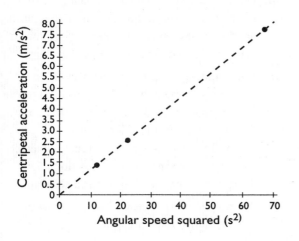

3. The units of the slope are meters. The slope represents the distance the accelerometer was from the center of the turntable.

4. Graphs will vary depending on data. A graph based on the sample data is shown below. The graph suggests that there is a linear relationship between radius and centripetal acceleration.

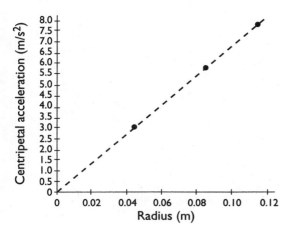

5. The units of the slope are s^{-2}. The slope for the sample data is $68.1\ s^{-2}$. The relationship of this slope to the physical parameters of the experiment is not obvious—the square root of the slope corresponds to the angular speed of 8.16 rad/s.

ANSWERS TO CONCLUSIONS

6. The centripetal force is directed inward. The accelerometer showed a positive value when the arrow was pointed toward the center of the turntable.

7. The graphs and data suggest that the centripetal acceleration is equal to the radius times the angular speed squared, or

$$a_c = r\omega^2$$

8. In order for the centripetal acceleration to be the same on the new ride as on the old ride, the angular velocity would have to be greater on the new ride by a factor of $\sqrt{2}$. If the new ride had the same angular velocity as the old ride, the centripetal acceleration would be half the centripetal acceleration on the old ride.

9. The centripetal acceleration would be 29 m/s^2 (about $3g$). A 62 kg rider would apply nearly 1800 N of force to the seat, directed away from the center of rotation. This does not include the additional downward force due to the weight of the rider.

10. The minimum acceleration required is 28 m/s^2 (this is independent of the rider's mass). If the ride has a radius of 2.0 m, it must have an angular velocity of at least 25 rpm in order to produce an acceleration of 28 m/s^2.

EXTENSIONS

1. The penny will leave a path going straight out from the center to the edge of the turntable. Make sure students do not use objects other than coins without your approval.

2. The path of the marble will vary depending on the magnitude and direction of its initial velocity. The rotation of the turntable will impart a force on the marble directed directly away from the center of the turntable.

3. This lab can be performed on a much bigger scale by using a playground merry-go-round. The experimental design can be similar; graphs showing the same relationships can be generated, and forces can be calculated based on the students' mass. Make sure to discuss safety issues with students. If necessary, obtain permission from parents to take students to the playground to perform the experiment.

The following procedures and suggestions should be provided to students for collecting data using a merry-go-round:

1. Use masking tape to mark the location of several radii along one of the pipes running outward from the hub of the merry-go-round. Select an increment that will give you four different radii for your particular merry-go-round. Record these radii in your data table.

2. Use a sample time interval of 0.50 s and 99 data points. This gives you 50 seconds to collect your data.

3. Get on the merry-go-round and spin it at a constant speed. Practice maintaining this speed even if the experimenter on the ride moves carefully inward or outward. Measure the time it takes for multiple rotations, then divide to determine the time for a single rotation. Record the period of rotation in the data table. This step can be completed while accelerations are being measured.

4. Hold the accelerometer at the first measured location, with the arrow horizontal and pointing toward the center of the merry-go-round. Start the data collection by pressing [TRIGGER] .

5. After holding the accelerometer in this position for 5–10 seconds, carefully move it to the second measured point, holding it there for a similar length of time. Continue moving to succeeding positions until you have collected data at all of your measured positions. Hold the speed of the merry-go-round constant.

6. Retrieve data from the CBL and display the acceleration versus time graph.

Technology Lab

Chapter 10

Newton's Law of Cooling

OBJECTIVES

In this lab, students will:

- **Develop** a model that can be used to test the insulating ability of various fabrics.

- **Predict** temperature changes and energy loss over time using the model.

- **Derive** a cooling constant for an insulating fabric under wet or dry conditions.

- **Compare** the cooling constants for different fabrics to determine the best fabrics for use in cold or wet conditions.

PLANNING

Recommended Time

1 lab period

Materials:

[for each lab group]

✔ graphing calculator with link cable

✔ CBL system

✔ PHYSICS application loaded in calculator

✔ TI or Vernier temperature probe

✔ 35-mm film canister with a hole in the lid

✔ samples of two or more insulating fabrics (wool, synthetic fleece, polypropylene, etc.)

✔ rubber bands

[for entire class]

✔ hot water

✔ pipette

✔ balance

Classroom Organization

This activity should be performed in groups of two or more.

Materials Preparation

You may want to set up lab stations beforehand. If you have more than one lab class, the equipment can be left standing from one class to the next.

This experiment calls for hot water at about 60°C. You may be able to get water this hot directly from the faucet. If your hot tap water is not hot enough, you may wish to heat a large beaker of water using a hot plate set on low.

Film canisters can often be obtained for free where film is developed. Carefully punch holes in the lids just large enough to allow temperature probes to be inserted.

Fabric samples can be obtained by purchasing a remnant from a fabric store. Only very small quantities are needed to cover a film canister. Use a lightweight wool fabric for the "natural" fabric. For the synthetic fabric, use a piece of synthetic fleece or polypropylene. Army surplus stores often sell inexpensive wool and polypropylene clothing that may be cut into samples. A 2 inch by 4 inch rectangular piece of fabric fits nicely around a film canister.

Techniques to Demonstrate

You may want to explain to students that Newton's "law" of cooling is a mathematical model that describes the way things cool fairly accurately. It is not a law in the same sense as Newton's laws of motion.

ANSWERS TO DEVELOPING THE MODEL

1. When t is 0, e^{-kt} is equal to 1, so T_{diff} is equal to T_0.

2. When t becomes very large, e^{-kt} approaches zero, so T_{diff} also approaches zero.

3. A large k implies that the warm body will cool relatively quickly. A good insulator has a small cooling constant, k.

4. Because the temperature of the body remains constant, it is impossible to use temperature changes to measure the rate of energy transfer.

CHECKPOINTS

Step 6: Inform students whether they should use hot tap water or water heated using a hot plate.

Step 7: If students measure an exact quantity of water, the k values between groups will be comparable. 25 grams of water will fill a black film canister approximately 3/4 full. If you choose to use different containers, you may have to modify this quantity.

Step 10: You may want to prepare another activity or reading for students to do while waiting the 20 minutes while the CBL is collecting data.

POST-LAB

Cleanup and Disposal

Very little cleanup is required. If the experiment is not to be performed again, store all equipment securely.

DATA TABLES WITH SAMPLE DATA

Fabric type	wool
Wet or dry?	dry
Room temperature (°C)	22.4

A	34.2
B	5.0×10^{-4}
Initial temperature difference, T_0	34.2°C
Cooling constant, k	5.0×10^{-4} s^{-1}

Fabric and conditions of test	Cooling constant, *k*
dry wool	5.0×10^{-4} s^{-1}
wet wool	7.8×10^{-4} s^{-1}
dry synthetic fleece	5.3×10^{-4} s^{-1}
wet synthetic fleece	6.1×10^{-4} s^{-1}

ANSWERS TO ANALYSIS

1. This step subtracts the room temperature from the data in the calculator list L_2.

2. A graph of the sample data with an exponential curve fit is shown below.

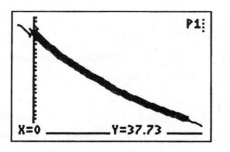

3. The value of T_0 is equal to the value of A in the fitted curve, with units of °C. The value of k is equal to the value of B in the fitted curve, with units of s^{-1}.

ANSWERS TO CONCLUSIONS

1. It appears that Newton's exponential model is a very good fit for this data.

2. The graph suggests that as the temperature of the water gets closer to the air temperature, the rate of energy transfer slows down.

3. For this sample data $T_{diff} = 34.2e^{-(5.0 \times 10^{-4})t}$. Therefore, after 60 minutes (3600 s), the temperature difference is 5.65°C. After 10 hours (36 000 s), the temperature difference is 5.2×10^{-7} °C. According to the model, the temperature of the water will never completely reach the air temperature.

4. Starting this experiment with a smaller or larger temperature difference should not affect the value of k. The data points would have fallen on a different place on the graph of the same function.

5. Sample values of k are shown in the third table in the Data Tables with Sample Data. Prepare a table like this for student groups to report their data on the board.

6. Answers will vary. For the sample data, it appears that wool has a small advantage over synthetic fleece when both are dry, but the fleece has a better insulating ability under wet conditions. Students should choose to recommend the fabric that has the lowest value of k when wet. For the sample data, this is the synthetic fleece.

EXTENSIONS

1. When you lower the temperature from 68°F to 65°F, the temperature difference between the inside and the outside of the house is less. Therefore, the rate of energy loss is reduced. Because the rate of energy loss is reduced, then the rate of energy production required to replace the lost energy would be less.

2. Following Newton's law of cooling, the cup stays hotter if the cream is added first. The early addition of the cream reduces the difference in temperature between the coffee and the air so it reduces the rate of energy loss during the waiting period.

3. Experiment designs should involve measuring some quantity over time, then fitting an exponential curve to the data.

Technology Lab Chapter **12**

Pendulum Periods

OBJECTIVES

In this lab, students will:

- **Design** experiments to determine the factors that may affect the period of a pendulum.

- **Test** the relationship between length, mass, amplitude and the period of a pendulum.

- **Recommend** the specific characteristics of clock pendulums using their data.

PLANNING

Recommended Time

1 lab period

Materials

[**for each lab group**]

- ✔ graphing calculator with link cable
- ✔ CBL system
- ✔ PHYSICS application loaded in calculator
- ✔ Vernier photogate and CBL adapter cable
- ✔ 2 support stands
- ✔ table clamp, pendulum clamp, and right-angle clamp
- ✔ masses of 100, 200, and 300 g with hooks
- ✔ meterstick
- ✔ protractor
- ✔ graph paper

Classroom Organization

This activity should be performed in groups of three.

Techniques to Demonstrate

Show students how to launch the photogate so that it swings in a straight line. This will help students avoid damaging the photogate.

Materials Preparation

You may want to set up lab stations beforehand. The equipment can be left standing from one class to the next.

A two-string support will confine the pendulum swing to a lane, avoiding photogate damage. Make sure strings are tied securely to the rods so they will not slide.

Be sure that the support stands holding the pendulums are rigid. If a support stand flexes with the pendulum swing, the period will be altered.

> ↗ internet **connect**
>
> go.
> hrw
> .com
>
> **GO TO:** go.hrw.com
> **KEYWORD:** HF2 TECHLABS
> This lab may also be performed with TI CBL 2 or Vernier LabPro systems. Instructions are available at this Web site.

ANSWERS TO DEVELOPING THE MODEL

1. For five periods, the pendulum will make 11 passes through the photogate.

2. A period is defined as a complete swing, including over and back. Therefore, to determine the period, the elapsed time between two passes must be measured.

3, 4. Answers will vary. Students should compare their predictions to their answers in the Conclusions after completing the lab.

CHECKPOINTS

Step 8 If students are not familiar with the operation of the photogate, encourage them to play with the photogate by passing a finger through the gate to see how the CBL and the PHYSICS application respond.

Step 9 A good alternate strategy for measuring the period at a particular amplitude is to start the pendulum at a little more than the desired amplitude and let the swing die down slightly before making the period measurement.

POST-LAB

Cleanup and Disposal

Very little cleanup is required. If the experiment is not to be performed again, store all equipment securely.

DATA TABLES WITH SAMPLE DATA

Part I Amplitude

Amplitude (°)	Average period, T (s)
2	1.688
5	1.691
10	1.695
20	1.702
30	1.717

Part II Length

Length, l (cm)	Average period, T (s)
50	1.434
60	1.567
70	1.680
80	1.789
90	1.902
100	2.008

Part III Mass

Mass, m (g)	Average period, T (s)
100	1.449
200	1.433
300	1.426

ANSWERS TO ANALYSIS

1. A graph of period versus amplitude for sample data is shown below. The data suggest that a change in amplitude causes only very small changes in the period.

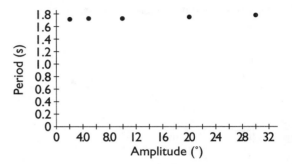

2. A graph of period versus length for sample data is shown below. The data suggest that there is a significant relationship between length and period.

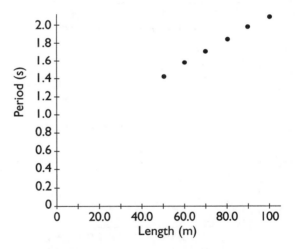

3. A graph of period versus mass for sample data is shown below. The data suggest that there is little or no relationship between mass and period. To confirm this, more data points would need to be collected.

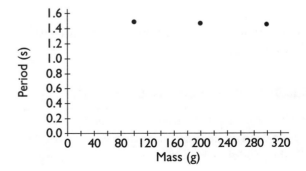

4. Two graphs based on sample data are shown below. The graph of T versus l^2 may appear to show a linear relationship, but there is no direct proportion; the fitted line does not go through the origin. The graph of T versus the square root of l appears to be a direct proportion, with the line fitting the data closely and passing through the origin.

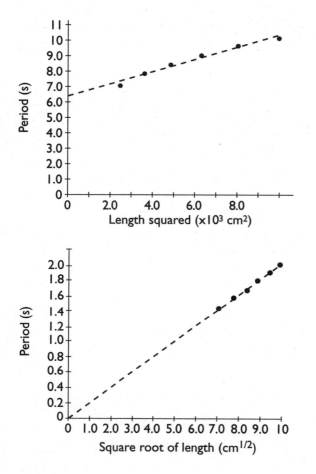

5. The following equation summarizes the relationship between period and string length for a pendulum:

$$T = k\sqrt{l}$$

The sample data yield a proportionality constant of 0.201 cm/s$^{(1/2)}$.

ANSWERS TO CONCLUSIONS

6. The antique clock needs a pendulum with a period of 1.875 seconds. To have this period, the pendulum should be 87 centimeters long.

7. Design the pendulum so the weight can be lowered or raised.

8. In the pendulum used in the experiment, the mass of the string is negligible. On a clock, the arm of the pendulum itself has mass, so the center of mass is closer to the pivot point.

9. The center of mass of the pendulum should be about 87 centimeters from the point where the pendulum attaches. The pendulum should swing through a narrow arc ($< 10°$). The mass of the pendulum should be great enough that air resistance and friction are negligible. Finally, the weight should be able to be moved up and down.

EXTENSIONS

1. Temperature changes may cause the pendulum to expand or contract, which could change the effective length and therefore the period of the pendulum. Other considerations may include humidity or wind.

2. A college physics textbook may provide the following equation to model the relationship between the amplitude and period of a pendulum:

$$T = 2\pi\sqrt{\frac{l}{g}}\left(1 + \frac{1}{4}\sin^2\frac{\theta}{2} + \frac{9}{64}\sin^4\frac{\theta}{2} + \ldots\right)$$

For small amplitudes, the series in parentheses is very nearly equal to one. As the amplitude increases, the period begins to increase significantly.

3. There are several ways to use the data to calculate the value of g. The following method uses the proportionality constant of 0.201 s/cm$^{(1/2)}$ to find g.

$$T = 2\pi\sqrt{\frac{l}{g}}$$

$$0.201 \text{ s/cm}^{(1/2)} = \frac{2\pi}{\sqrt{g}}$$

$$g = \left(\frac{-2\pi}{0.201}\right)^2 \frac{\text{cm}}{\text{s}^2} = 977.2 \text{ cm/s}^2$$

Technology Lab

Chapter 13

Sound Waves and Beats

OBJECTIVES

In this lab, students will:

- **Develop** a setup that allows them to analyze sound waves for frequency and amplitude.

- **Analyze** the waveforms of a tuning fork and a musical instrument.

- **Model** waveforms using a mathematical expression derived from their data.

- **Observe** beat frequencies when two waveforms are combined.

- **Evaluate** their ability to tune by ear based on beat frequencies in the combined waveform.

PLANNING

Recommended Time

1 lab period

Materials

[for each lab group]

✔ graphing calculator with link cable

✔ CBL system

✔ PHYSICS application loaded in calculator

✔ Vernier microphone

✔ tuning fork

✔ rubber striker for tuning fork

✔ stringed instrument (or other musical instrument that may be tuned)

Classroom Organization

This activity should be performed in groups of two or more.

Materials Preparation

You may want to set up lab stations beforehand. If you have more than one lab class, the equipment can be left standing from one class to the next.

Use quality tuning forks with large tines. Smaller tuning forks can be used, but they must be held very close to the microphone. When using tuning forks, strike them against a soft object, such as a rubber mallet, rubber sole, or a rubber stopper. Special rubber strikers for tuning forks are also available.

If you don't have tuning forks, a pure electronic tone will also work. If you use an electronic keyboard, use a voice that produces as pure a sine wave as possible. On some keyboards, this is the flute sound.

Remind students several days before the scheduled lab activity that they will need to bring in musical instruments. If you do not have enough instruments, you may be able to borrow some from your school's music teacher. Any stringed instruments will work, even inexpensive ones.

It is important to use instruments for which the pitch of an open string or an easily fretted note will correspond to the pitch of the tuning fork. For an A tuning fork (about 440 Hz), a guitar (third string, second fret), a banjo (fifth string can be tuned up), a violin (second string), a mandolin (second pair of strings), and a dulcimer (tuning varies) will all work well.

internet**connect**

GO TO: go.hrw.com
KEYWORD: HF2 TECHLABS
This lab may also be performed with TI CBL 2 or Vernier LabPro systems. Instructions are available at this Web site.

Techniques to Demonstrate

Show students how to safely strike a tuning fork against a rubber striker or stopper. Caution students not to strike tuning forks against hard, rigid surfaces. Doing so may damage the tuning forks. In addition, the tones produced will be impure, with lots of unwanted harmonics.

Instruct students on the proper handling of musical instruments. Warn students not to overtighten strings.

DATA TABLES WITH SAMPLE DATA

The following sample data were generated with a 440 A tuning fork and a banjo with the fifth string tuned to the tuning fork.

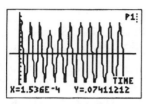

A sample waveform for a 440 Hz A tuning fork.

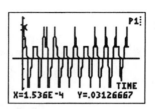

A sample waveform for a banjo tuned to the 440 Hz A tuning fork.

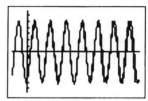

The tuning fork waveform superimposed with the model.

Procedure

Source of sound	Number of cycles	First maximum (s)	Last maximum (s)	Δt (s)	Period (s)	Calculated frequency (Hz)
tuning fork	4	.00395	.01292	0.00896	0.00224	446
tuned string	4	0.00130	0.00960	0.00420	0.00210	456

Source of sound	Peak (V)	Trough (V)	Amplitude (V)
tuning fork	0.037	−0.031	0.034
tuned string	0.0312	−0.0314	0.0313

Source of sound	Amplitude parameter A (V)	Frequency parameter B (s^{-1})	$f = B / 2\pi$ (Hz)
tuning fork	0.035	2804	446.3
tuned string	0.033	2868	456.5

Analysis

Number of cycles	First maximum (s)	Last maximum (s)	Δt (s)	Beat period (s)	Beat frequency (Hz)
3	0.00355	0.2977	0.2941	0.0980	10.2

ANSWERS TO DEVELOPING THE MODEL

1. A peak on the graph should be labeled as the maximum compression. A trough should be labeled as the maximum rarefaction.

2. The amplitude of the model represents half the difference between the peak compression and the peak rarefaction. The volume of a sound is determined by the vertical difference between these two points. The larger the volume of the sound, the larger the amplitude on the graph of the sound.

3. The period of the wave shown is 0.0035 s.

4. The frequency is the inverse of the period. For this wave, the frequency is 286 Hz.

5. As the value of A goes up and down, the graph will stretch or shrink in the vertical direction. As the value of f increases, the sine curve will be compressed horizontally. As f decreases, the sine curve will be stretched horizontally.

Checkpoints

Step 1: Make sure students put their calculators in radian mode. This is necessary for the PHYSICS application to properly interpret waveform data from the CBL.

Steps 6 and 16: Students may find that the peaks and troughs of the generated waveforms are not exactly the same distance from the y-axis. If this happens, suggest that they generate a new waveform. If they can't get one exactly balanced, they can try to "average" the peaks and troughs when generating the model waveform. Small differences in the peak and trough distances should not significantly affect their results.

POST-LAB

Cleanup and Disposal

Very little cleanup is required. If the experiment is not to be performed again, store equipment securely.

ANSWERS TO ANALYSIS

1. This step involves preparing the calculator for analysis.

2. This step involves preparing the calculator for analysis.

3. This waveform shown below was generated from sample data by adding the mathematical model of the tuning fork tone to the one produced by the banjo after tuning.

4. See the fourth sample data table for beat frequencies calculated from sample data.

5. The final combined model shows a sinusoidal waveform with modulating amplitude. The amplitude reaches a peak when both of the waves have a crest at the same moment in time. The amplitude reaches a minimum when one wave has a crest and the other has a trough.

ANSWERS TO CONCLUSIONS

6. See the sample data tables for values generated from sample data. The most accurate way to determine the frequency is to look at many cycles of the waveform generated by the sound rather than looking at the model. The model is an approximation of the true waveform that was generated by matching only a small portion of the graph.

7. Answers may vary. The model fit the waveforms for the sample data well. Both the waveforms and the model showed sinusoidal curves. The waveforms, however, were not pure sine waves. This was especially true for the banjo waveform.

8. For the sample data, the frequency of the beats observed in the combined model was 10.2 Hz. This frequency is equal to the difference in frequencies between the banjo string and the tuning fork.

9. If the string had been tuned exactly to the tuning fork, there would have been no beats observed. This is because the crests and troughs of the two waves would always remain in the same relative position and the amplitude of the combined wave would not change. In practice, it would be impossible to get the string exactly in tune with the tuning fork.

10. For the sample data, the tuned string differed from the tuning fork by more than 10 Hz. This would not be an acceptable difference for most musical purposes. The person that tuned the banjo should probably rely on an electronic tuner.

11. When two notes are played separately, it is impossible to hear beats. When two notes are played at the same time, even if a musician can't recognize a difference in pitch, he or she can listen for beats to recognize that the two strings are not tuned correctly. By continuing to adjust the pitches of strings until the beats go away, the musican can get the instrument in good tune.

EXTENSIONS

1. Beat frequencies should increase as the notes get farther apart.

2. Noise cancellation technologies use destructive interference to dampen sound volumes.

Technology Lab

Chapter 14

Polarization of Light

OBJECTIVES

In this lab, students will:

- **Develop** an experiment that tests the ability of polarized sunglasses to reduce glare.

- **Measure** the transmission of light through two polarizing filters as a function of the angle between their axes.

- **Measure** the transmission of reflected light (glare) through a pair of polarized sunglasses as a function of angle.

- **Evaluate** the effectiveness of polarized lenses in a consumer report.

PLANNING

Recommended Time

1 lab period

Materials

[for each lab group]

✔ graphing calculator with link cable

✔ CBL system

✔ PHYSICS application loaded in calculator

✔ TI or Vernier light sensor

✔ DC light source

✔ two polarizing filters with marked axes

✔ pair of polarized sunglasses

✔ smooth horizontal surface to reflect light

✔ cardboard screen

✔ three support stands with clamps for holding light source, light sensor, cardboard screen

✔ tape

✔ toothpick

✔ protractor

Classroom Organization

This activity should be performed in groups of two or more.

Limit the number of groups so that each group will have access to a table near a window that receives direct sunlight. You should plan to do this experiment when sunlight will be coming through the windows onto the tables. If this is not possible, Part II of this experiment can be done using an artificial light source, as in Part I.

Materials Preparation

You may want to set up lab stations beforehand. If you have more than one lab class, the equipment can be left standing from one class to the next.

internet connect

GO TO: go.hrw.com
KEYWORD: HF2 TECHLABS
This lab may also be performed with TI CBL 2 or Vernier LabPro systems. Instructions are available at this Web site.

Techniques to Demonstrate

You may want to give students unmarked polarizing filters and show them how to determine the axis of polarization using glare from a horizontal surface.

ANSWERS TO DEVELOPING THE MODEL

1. Reflected light offers a test under actual use conditions.

2. When the difference between the two axes is 0°, $\cos^2\theta = 1$. Therefore, $I = I_0$.

3. When the difference between the axes of polarization is 90°, $\cos^2\theta = 0$. Therefore, $I = 0$.

4.

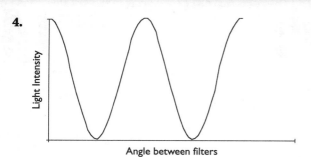

5. Darker lenses would reduce the total amount of light entering the eye; however, glare would still be intense unless very dark lenses were used. Very dark lenses would make it difficult to see objects that are not in high contrast to the background. The use of polarized lenses will reduce glare to nearly zero and still allow a great deal of light to enter the eyes.

CHECKPOINTS

Step 1: Make sure calculators are in degree mode.

Step 14: If tables in direct sunlight are not available, students may use the setup for Part I to do Part II, using the sunglasses instead of the second polarizing filter.

POST-LAB

Cleanup and Disposal

Very little cleanup is required. If the experiment is not to be performed again, store equipment securely.

DATA TABLE WITH SAMPLE DATA

Artificial light and two filters	
background light	0.03965
maximum intensity	0.15513

Reflected light and sunglasses	
background light	0.04852
maximum intensity	0.05275

ANSWERS TO ANALYSIS

1. Graphs based on sample data are shown below. The top graph shows light intensity versus angle for two polarized filters. The bottom graph shows light intensity versus angle for polarized sunglasses filtering reflected light. Both follow the general $\cos^2$ curve. The graph of the two filters shows a closer fit with Malus's model. Both graphs show maxima and minima alternating at 90° intervals.

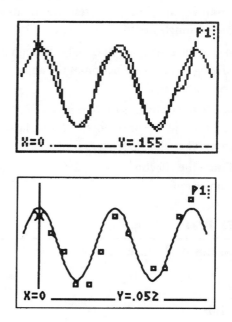

2. Both sets of data are consistent with Malus's model. The data for the sunglasses shows more variation, but this is understandable because it was harder to control all variables in this part of the experiment. The sample data were collected under low cloud conditions. During data collection, the amount of incident light changed slightly, probably due to a change in cloud thickness.

3. The difference between the two readings was greater with the two filters than with the sunglasses. A greater percentage of light was eliminated when using the two filters.

4. The reduction in light intensity for the sunglasses was 8 percent.

ANSWERS TO CONCLUSIONS

5. It is important that sunglasses allow some light to pass through. If sunglasses were to block out too much light, it would be difficult to see.

6. Making sunglasses darker would reduce all of the light coming into the eyes. The glare would still be the brightest light experienced by the wearer.

7. To reduce 99 percent of the glare from horizontal surfaces, polarized lenses must be oriented so that the axis of polarization is less than 1 degree from the vertical. However, even lenses aligned perfectly would not completely block glare from all surfaces, because not all surfaces are oriented the same. Furthermore, the axis of polarization of the lenses will change orientation when the wearer tilts his or her head.

8. Answers will vary. In general, even inexpensive polarized glasses will reduce glare considerably.

9. Answers will vary. Presentations should in some way use results from the experiment to recommend for or against the use of polarized sunglasses.

EXTENSIONS

1. The middle filter transmits only light parallel to its axis. As long as the middle filter axis is not perpendicular to the first filter axis, the light transmitted by the middle filter has a polarization component of the light transmitted by the first filter. The second filter rotates the polarization of the light, so some light makes it through the third and final filter.

2. Students should compare their own conclusions to those presented by others. If possible, students should investigate the methods used by others to evaluate the effectiveness of sunglasses.

3. Most polarized lenses use a polarizing film manufactured in a process pioneered by the Polaroid corporation in the 1930s. Complex crystals are aligned into parallel strata as the film material is stretched.

Technology Lab

Chapter 18

Capacitors

OBJECTIVES

In this lab, students will:

- **Model** a timing device using a circuit with a resistor and a capacitor.

- **Measure** the change in potential over time for a circuit containing a capacitor and resistor as the capacitor discharges.

- **Derive** the time constant of a resistor-capacitor circuit.

- **Apply** the results to a strobe flash system.

PLANNING

Recommended Time

1 lab period

Materials

[for each lab group]

✔ graphing calculator with link cable

✔ CBL system

✔ PHYSICS application loaded in calculator

✔ TI or Vernier voltage probe

✔ 10 μF nonpolarized capacitor

✔ 22 kΩ and 47 kΩ resistors

✔ 9 V battery with battery clip

✔ single-pole, double-throw switch

✔ connecting wires and clips

Classroom Organization

This activity should be performed in groups of three or more.

Materials Preparation

You may want to set up lab stations beforehand. If you have more than one lab class, the equipment can be left standing from one class to the next.

Resistors with a resistance of more than 50 kΩ should not be used in this experiment. Larger values will give incorrect results due to the voltage-dependent input impedance of the CBL.

If possible, prepare students by demonstrating a circuit using a 1 F capacitor and a flashlight bulb. Charge the capacitor, then allow it to discharge through the bulb. Have students observe the decrease in brightness over time. This demonstration will help prepare students for the CBL experiment, which takes place much more quickly.

The experiment calls for 10 μF, nonpolarized capacitors. Any voltage rating can be used, but capacitors that are rated for lower voltages are usually less expensive. If you choose to use an ordinary polarized electrolytic capacitor, the negative side of the capacitor must be connected to the negative battery terminal, with the resistor in between, or the circuit will not work. Electrolytic capacitors can be destroyed if they are wired backward.

The color code for resistor values and tolerances is explained in the Teacher's Notes for the Chapter 20 Technology Lab. Resistors are typically marked with a tolerance of 5 percent or 10 percent. Capacitors often have 10 percent or 20 percent tolerances. Students will probably not find agreement with the marked values.

🔾 internet **connect**

go.
hrw
.com

GO TO: go.hrw.com
KEYWORD: HF2 TECHLABS
This lab may also be performed with TI CBL 2 or Vernier LabPro systems. Instructions are available at this Web site.

Techniques to Demonstrate

You may want to perform a dimensional analysis to show students that the units of ohms multiplied by farads are seconds:

$$\text{ohm} \times \text{farad} = \left(\frac{\text{volt}}{\text{ampere}}\right) \times \left(\frac{\text{coulomb}}{\text{volt}}\right) = \left(\frac{\text{coulomb}}{\text{ampere}}\right)$$

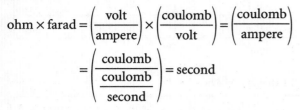

$$= \left(\frac{\text{coulomb}}{\frac{\text{coulomb}}{\text{second}}}\right) = \text{second}$$

ANSWERS TO DEVELOPING THE MODEL

1. With the switch in the current position, the capacitor will be charged. There is a complete circuit between the power supply and the capacitor.

2. When the switch is moved to the other position, the capacitor will discharge through the resistor.

3. According to the equation, an increase in C will result in a slower discharge. An increase in R will also result in a slower discharge. Therefore, to measure short time intervals, a large resistance should be used.

4. The larger the time constant, the slower the rate of discharge.

5. The specific capacitors and resistors used in the timer may vary from their labeled values, and the circuit itself will have some resistance, so making predictions based on the labeled values would not be accurate. It would be more accurate to use an actual circuit to collect calibration data.

CHECKPOINTS

Step 1: Other combinations for R and C can be used. If the time constant RC is significantly different from 0.5 s, you will need to adjust the data collection parameters.

Step 8: Students should take data for approximately three time constants and gather about 50 data points during that time. If data are collected for too long a time period, the potential difference will approach zero, and the exponential fit will be poor.

Step 13: The exponential fit done by the PHYSICS application uses the equation $y = Ae^{-Bx}$. This is different from the exponential fit to $y = ab^x$ normally used by TI calculators. Although the same information is obtained, the base-e form was chosen for this experiment so that the fit parameters would correspond directly to terms in the model equations.

POST-LAB

Cleanup and Disposal

Very little cleanup is required. If the experiment is not to be performed again, store equipment safely.

DATA TABLES WITH SAMPLE DATA

Resistance, $R(k\Omega)$	Capacitance, $C(\mu F)$	Time constant, $RC(s)$	Curve fit parameters			Average of 1/B
			A	**B**	**1/B**	
47	10		8.10	2.06	0.485	
47	10	0.47	8.02	2.01	0.498	0.486
47	10		8.15	2.10	0.476	
22	10		8.05	4.17	0.240	
22	10	0.22	8.12	4.11	0.243	0.240
22	10		8.03	4.20	0.238	

Flash duration (s)	Resistor choice (kΩ)	Initial potential (V)	Final potential (V)
0.01	22	8.05	8.03
0.05	22	8.05	7.95
0.1	47	8.1	7.72
0.25	47	8.1	7.17

ANSWERS TO ANALYSIS

1. See the first sample data table.

2. *A* represents the initial potential difference at time equal to zero.

3. *B* corresponds to the inverse of the time constant; that is, $1/B$ is equal to the time constant.

4. See the first sample data table.

5. See the first sample data table.

6. For sample data, the calculated time constant value and the value of $1/B$ are within one percent of each other. Based on this result, it is probable that both the resistor and the capacitor are within the 5 percent tolerance suggested by the manufacturer.

7. The smaller resistor resulted in faster discharge times.

ANSWERS TO CONCLUSIONS

8. There is minimal variation between trials. These components could work reliably in a timer.

9. The 47 kΩ resistor could be used for timing events up to about 0.25 seconds. The 22 kΩ resistor could be used for timing events up to about 0.1 seconds. See the second data table with sample data (above).

EXTENSIONS

1. When a plot is made by graphing the natural log of the potential difference against time, the result is a straight line with a slope equal to $-1/RC$.

2. After one time constant elapses, approximately 37 percent of the potential remains. After two time constants, the potential difference is reduced to 13 percent of the initial value. After three, the potential difference is down to 5 percent of the initial value.

3. The light bulb is not ohmic; that is, the current is not proportional to the potential across it. As the current starts to flow through the wires, the bulb's filament is cool. The wires warm up quickly and their resistance increases. As a result, a graph of potential difference plotted against time does not correspond to a simple exponential equation.

4. Different values for *R* and *C* will create different time constants. However, if the product *RC* is the same, the time constant will be the same.

5. Capacitors in parallel behave as a single capacitor with the sum of the individual capacitances. The time constant will be approximately doubled.

6. Capacitors in series behave as a single capacitor with half the capacitance of one of the individual capacitors. The time constant will be approximately half.

Technology Lab

Chapter 19

Electrical Energy

OBJECTIVES

In this lab, students will:

- **Model** a toy car and evaluate the efficiency of an electric motor under different loads.

- **Measure** the power and electrical energy used by an electric motor under different loads.

- **Calculate** the efficiency of the motor using the data collected.

- **Evaluate** the motor efficiency and make recommendations regarding design modifications.

PLANNING

Recommended Time

1 lab period

Materials

[for each lab group]

✔ graphing calculator with link cable

✔ CBL system

✔ PHYSICS application loaded in calculator

✔ Vernier current and voltage probe system

✔ 2 Vernier CBL adapter cables

✔ adjustable-voltage DC power supply

✔ electric motor

✔ mass set

✔ string

✔ table clamp and rod

✔ utility clamp

✔ short dowel rod or thread spool

✔ wires

✔ clips to hold wires

Classroom Organization

This activity should be performed in groups of two or more.

Materials Preparation

You may want to set up lab stations beforehand. If you have more than one lab class, the equipment can be left standing from one class to the next.

The procedure calls for a 12 V DC motor. This lab works well with smaller motors also. Make sure students know the voltage rating of the motors they are using and that they do not exceed that voltage during the experiment. Students may need to use different masses if they use a smaller or larger motor.

If a motor does not have a long shaft or spindle for wrapping the string, there are several ways to make one. Probably the easiest way is to carefully drill a hole in the end of a wooden dowel rod (3/8″ × 2″ or so) and slip it onto the shaft of the motor. An empty thread spool also works well.

Instead of a table clamp, you may use a support stand clamped securely to the table.

Mount motors below head level to reduce the risk of injury.

internet connect

GO TO: go.hrw.com
KEYWORD: HF2 TECHLABS
This lab may also be performed with TI CBL 2 or Vernier LabPro systems. Instructions are available at this Web site.

Techniques to Demonstrate

Demonstrate setting up the spindle and the motor so that the string will wind around the axle as the motor turns.

ANSWERS TO DEVELOPING THE MODEL

1. To determine the amount of potential energy gained by the mass being lifted by the motor, you would need to measure the mass being lifted and the distance the mass is lifted. In addition, you would need to know the value of *g*, the acceleration due to gravity.

2. The mathematical relationship between potential difference, current, and power is shown in the following equation:

   power = potential difference × current

3. W·s is the same as joules, the units for energy. When energy is used at the rate of 1 joule per second, it is the same as using power at the rate of 1 watt.

4. If there were no mass attached to the motor, the calculated change in gravitational potential energy would be equal to zero. Therefore, it would be impossible to measure the efficiency of the motor using this model.

5. If the load placed on the motor was very high, the motor would not be able to lift it and the work completed by the motor would be calculated as 0 joules. The efficiency of the motor would then be calculated as 0 percent.

6. As the mass is lifted, its gravitational potential energy increases. At the same time, electrical energy is being expended by the motor. While some of the electrical energy is used to lift the mass, some of it is used to overcome resistance and friction in the motor. Ultimately, the energy lost to resistance and friction is transferred as heat.

7. Answers will vary. Most students will overestimate the efficiency of the electric motor.

CHECKPOINTS

Step 2: Check students' circuits to make sure they match **Figure 19-2** in the Procedure.

Step 6: Be sure to have students check the voltage rating on the motors they are testing. Warn students not to turn up the voltage above the voltage rating of the motor. Voltages that are too high can burn out the motors.

Step 8: Check strings for wear, especially if they have been tangled.

Step 9: Caution students against using the motors to lift the masses too quickly. Make sure students wear safety glasses and lift the masses at controlled speeds.

Step 14: The efficiency will decrease if the motor is overloaded. At this point, a slight push can be used to start the motor if it fails to lift the load initially.

DATA TABLES WITH SAMPLE DATA

Distance mass was lifted (m)	0.97 m
Voltage rating of motor (V)	12 V

Trial	Load lifted (kg)	Force exerted (N)	Electrical energy used (J)	Mechanical energy output (J)	Efficiency (%)
1	0.010	0.098	0.958	0.098	10.2
2	0.020	0.196	1.79	0.196	10.9
3	0.030	0.294	2.97	0.294	9.9
4	0.040	0.392	unable to lift	0	0

POST-LAB

Cleanup and Disposal

Very little cleanup is required. If the experiment is not to be performed again, store equipment safely. Check strings for wear and discard any strings that are frayed, stretched, or worn.

ANSWERS TO ANALYSIS

1. See sample data tables.

2. See sample data tables.

3. See sample data tables.

4. When a 40 g mass was attached to the motor, the motor was unable to lift it. Even though there was a potential difference across the motor, and current in it, the mechanical work output was zero.

ANSWERS TO CONCLUSIONS

5. The electrical energy that went into the motor was converted to thermal energy, sound, and light, in addition to mechanical energy.

6. Answers will vary. Based on sample data, the highest efficiency that AutoRemote can expect is about 11 percent.

7. A drag force of about 0.2 N results in the highest efficiency. A drag force of 0 N would result in the longest-lasting batteries.

8. Answers will vary, but there are three general approaches that can be used to lengthen the life of the battery. One is to reduce the overall weight of the car. A second is to reduce drag by using better quality wheels and other moving parts on the car. The third is to replace the motor with one that has significantly higher efficiency. Reducing the weight of the car may not increase the price, but the car may be less sturdy if the weight is reduced. Using better wheels and parts would probably increase the price. Using a more efficient motor would probably increase the price.

EXTENSIONS

1. Students' graphs should show the efficiency of the motor plotted against the mass lifted.

2. Check students' setups before they proceed. To perform this as a class activity, have different groups perform trials at different speeds, then have the groups share their results.

3. Research motor design to determine how design changes may affect motor efficiency.

4. To do this, students will need to time the mass as it rises. They can then divide the change in gravitational potential energy in joules by the time it took to raise the mass to get the power in watts. One horsepower is equal to 746 watts.

5. Students should use a padded catch box to catch the falling mass. Because the mass will fall relatively quickly, students may need to perform several trials to obtain good data. Check setups and circuits before students proceed. Make sure masses are firmly attached and that the motor is securely clamped to the edge of the table.

Technology Lab

Chapter 20

Series and Parallel Circuits

OBJECTIVES

In this lab, students will:

- **Develop** a model for measuring currents in both parallel and series circuits.

- **Measure and analyze** voltage and current in standard series and parallel circuits designed with two resistors.

- **Design** a series of circuits using a combination of series and parallel resistor elements to achieve specific currents.

- **Evaluate** the circuit design strategies of Current Design's new line of instrument pickups and microphones.

PLANNING

Recommended Time

1 lab period

Materials

[for each lab group]

- ✔ graphing calculator with link cable
- ✔ CBL system
- ✔ PHYSICS application loaded in calculator
- ✔ Vernier current and voltage probe system (including dual-channel amplifier)
- ✔ two Vernier adapter cables
- ✔ low-voltage DC power supply
- ✔ two 50 Ω resistors
- ✔ two 68 Ω resistors
- ✔ momentary-contact switch
- ✔ connecting wires or breadboard

Classroom Organization

This activity should be performed in groups of two or more.

Materials Preparation

You may want to set up lab stations beforehand. If you have more than one lab class, the equipment can be left standing from one class to the next.

The low-voltage DC can be either from batteries or from a DC power supply. The procedure calls for 3 V. If the voltage used is different, the answers will vary from those provided below. The output voltage should be constant so that students can compare measurements made on different runs.

This experiment was designed using power resistors that will work for up to a 5 V power supply. If you use a lower voltage power supply, you can use resistors rated at lower power.

The recommended tolerance of the 50 Ω resistor is ±10%. The recommended tolerance of the 68 Ω resistor is ±5%.

∕ internet connect

go.
hrw.
.com

GO TO: go.hrw.com
KEYWORD: HF2 TECHLABS
This lab may also be performed with TI CBL 2 or Vernier Lab Pro systems. Instructions are available at this Web site.

Techniques to Demonstrate

Warn students that improperly connected circuits can lead to excessive heating. Show students how to handle circuit elements that may be hot. You may want to introduce your students to resistor codes as part of this lab. The color codes used for carbon resistors are shown in a table on the next page.

black	0	blue	6
brown	1	violet	7
red	2	gray	8
orange	3	white	9
yellow	4	gold	0.1
green	5	silver	0.01

Most ordinary carbon-composition resistors are labeled with four color bands. The bands are read in order from the end of the resistor with a color band nearest the end. The first two bands are read as a two-digit number. The third band is read as a power of 10 multiplier of that number.

$$\text{resistance} = (\text{2-digit number from first 2 bands}) \times 10^{(\text{third band digit})}$$

The fourth band is an indicator of the tolerance of the resistor as a percentage of the labeled value. The color code for the tolerance is as follows:

no color	20%	gold	5%
silver	10%	brown	1%

ANSWERS TO DEVELOPING THE MODEL

1. Answers will vary depending on the resistors used. The resistance of a 50 Ω resistor with a 10 percent tolerance could be anywhere in the range of 45 to 55 Ω. The resistance of a 68 Ω resistor with a 5 percent tolerance could be anywhere in the range of 64 to 71 Ω.

2. As each bulb was added in series, the resistance would go up, the total current would go down, and the bulbs would get dimmer. As each new bulb was added in parallel, the total current flow would increase. The bulbs would remain bright as each new bulb was added.

3. To determine the resistance of a light bulb or any other device, you could measure the current and the voltage across the device and then calculate the resistance using Ohm's law ($R = \Delta V/I$).

4. It would be important to make measurements on the actual circuit because resistors vary in

their tolerances. In addition, other components of the circuit may contribute additional resistance that is not easily factored into calculations.

CHECKPOINTS

Steps 7, 14, and 22: Before students close the switch, check their circuits to make sure they are connected properly. Make sure that current probes are in series and voltage probes are in parallel. Whenever students rewire a circuit, check the new configuration before they complete the circuit.

POST-LAB

Cleanup and Disposal

Very little cleanup is required. If the experiment is not to be performed again, store equipment safely. Be careful handling resistors; they may be hot.

ANSWERS TO ANALYSIS

1. $\Delta V_1 + \Delta V_2$ is approximately equal to ΔV_{tot}.

2. See the data table with sample data for Part I. For a series circuit, R_{eq} is the sum of the individual resistances.

3. See the data table with sample data for Part II. For a parallel circuit, R_{eq} is less than either of the individual resistances. Most textbooks provide the formula $\frac{1}{R_{tot}} = \frac{1}{R_1} + \frac{1}{R_2}$ for calculating the equivalent resistance of a parallel circuit.

4. As you add more resistors in series, the current flow decreases.

5. As you add more resistors in parallel, the current flow increases.

6. See the data table with sample data for Part III.

ANSWERS TO CONCLUSIONS

7. Answers will vary. For sample data, the percent difference between R_{eq} and the expected resistance values for the three circuits were 3.4, 2.6, and 1.5 percent, respectively. These were within the expected tolerance ratings of the resistors.

DATA TABLES WITH SAMPLE DATA

Part I

R_1 (Ω)	R_2 (Ω)	I (A)	ΔV_1 (V)	ΔV_2 (V)	R_{eq} (Ω)	ΔV_{tot} (V)
50	50	0.030	1.560	1.515	103.4	3.103
50	68	0.026	1.315	1.789	121.1	3.104
68	68	0.023	1.555	1.555	138.0	3.105

Part II

R_1 (Ω)	R_2 (Ω)	I (A)	ΔV_1 (V)	ΔV_2 (V)	R_{eq} (Ω)	ΔV_{tot} (V)
50	50	0.120	3.082	3.079	25.7	3.085
50	68	0.104	3.085	3.074	29.7	3.085
68	68	0.090	3.084	3.086	34.3	3.089

Part III

Circuit	Desired current (A)	Circuit description	Measured current (A)	percent deviation (%)
1	0.017	50 Ω, 50 Ω, and 68 Ω in series	0.019	12
2	0.013	50 Ω, 68 Ω, and 68 Ω in series	0.014	8
3	0.032	two 50 Ω in parallel; in series with 68 Ω	0.031	3
4	0.330	two 50 Ω and two 68 Ω, all in parallel	0.325	2

8. To lower the current flow, you could also decrease the power supply voltage.

9. Answers will vary. For sample data, Circuits 1 and 2 exceed the tolerance rating specified by the engineer. Circuits 3 and 4 work much better. Students should explain whether their circuits would be acceptable using data from their experiments.

10. It would be possible to drop the current as low as you wished simply by adding more resistors in series. If you had a power supply with enough power rating, you could increase the current by arranging the resistors in parallel.

EXTENSIONS

1. Answers will vary. You may wish to have students calculate these values as part of the experiment.

2. Answers will vary. One way to do this is to wrap a long piece of wire lengthwise around a piece of PVC pipe. Scrape off the insulation at 5 ft intervals. Use the current and voltage probe system to make measurements and calculate R_{eq} per foot. If students are going to perform their experiments, warn them to be careful of excessive heat in the wires, especially nichrome wires.

3. You may want to have students use light bulbs instead of resistors for the entire lab.

Technology Lab

Chapter 21

Magnetic Field Strength

OBJECTIVES

In this lab, students will:

- **Model** a current meter that uses a magnetic field sensor to measure current.

- **Investigate** the relationship between the magnetic field and the number of turns in a coil.

- **Investigate** the relationship between the magnetic field and the current in a coil.

- **Determine** the current in a low-voltage DC appliance using your model.

PLANNING

Recommended Time

1 lab period

Materials

[for each lab group]

- ✔ graphing calculator with link cable
- ✔ CBL system
- ✔ PHYSICS application loaded in calculator
- ✔ Vernier magnetic field sensor and CBL adapter cable
- ✔ TI or Vernier voltage probe
- ✔ 1 Ω power resistor
- ✔ long spool of insulated wire (at least 12 m)
- ✔ low-voltage DC motor, light bulb, or buzzer
- ✔ magnetic compass
- ✔ momentary-contact switch
- ✔ small square or circular frame or shoe box

Classroom Organization

This activity should be performed in groups of two or more.

Materials Preparation

You may want to set up lab stations beforehand. If you have more than one lab class, the equipment can be left standing from one class to the next.

The frame should be about 30 cm on a side or in diameter. A temporary frame can be built from a sturdy cardboard box, or you can construct a plastic or wooden frame. Do not use materials containing iron. A frame can also be made from a piece of PVC pipe or mailing tube. Drill a hole in the side just large enough for the magnetic field sensor to be inserted. Using tubing allows students to add loops to the coil more easily.

A current of 3 A is required to perform this experiment. For best results and to protect the power supply, use a current-controlled power supply. Another way to limit the current is to use a power resistor in series with the coil.

The 1 Ω resistor must dissipate at least 5 W.

A momentary-contact switch is recommended because some power supplies may become hot if operated continuously.

⚡ internet connect

GO TO: go.hrw.com
KEYWORD: HF2 TECHLABS
This lab may also be performed with TI CBL 2 or Vernier LabPro systems. Instructions are available at this Web site.

Techniques to Demonstrate

You may want to show students how the position of the magnetic field sensor can affect the reading of magnetic field strength. Some students may become confused regarding the importance of orienting the sensor in the correct position to get readings. You may wish to have students explore the behavior of the sensor by rotating it near a bar magnet.

ANSWERS TO DEVELOPING THE MODEL

1. You could use Ohm's law to calculate the current from the voltage and the resistance. When the resistance is 1 Ω, the current in amperes will be equal in magnitude to the potential difference in volts.

2. One way to find the correct position of the sensor is to monitor the measurement while moving the sensor into various positions. To find the north pole of the magnetic field induced by the coil, you can use the right hand rule or a magnetic compass.

3. The relationship can be described mathematically as an equation for a line. The current can be determined by rewriting the equation so that the current is equal to the strength of the magnetic field divided by the slope.

4. Earth's magnetic field as well as currents or metal objects nearby will affect the background magnetic field. These fields may change during the course of the experiment and may affect the results unless they are subtracted from the data by zeroing the instrument.

CHECKPOINTS

Step 12: In Part II, the current is set to 3 A at the start and is not monitored during data collection. In order to keep the current at 3 A while removing coils from the frame, make sure that students do not change the overall length of the wire. Have them remove a coil from the frame and leave the excess wire as connecting cable between the power supply and the coil.

Step 14: In each part of this experiment, students are asked to zero the sensor before collecting data. This is an important step because the local magnetic field can change over short periods of time.

Step 15: It is essential that the sensor remain in the same position during a data run. A typical classroom has many sources of magnetic fields. If the sensor is moved even a few centimeters, the local field can be significantly different. This is not a malfunction of the magnetic field sensor, but is due to magnetic fields being ubiquitous and uneven.

Step 16: Make sure the switch is closed when unwinding the coils.

Step 25: Check students' circuits. Voltage levels should be appropriate for the device added to the circuit.

POST-LAB

Cleanup and Disposal

Very little cleanup is required. If the experiment is not to be performed again, store equipment securely.

ANSWERS TO ANALYSIS

1. There is a linear relationship between the magnetic field strength and the number of turns.

2. Equations should have the form $y = Ax + B$. The y-intercept corresponds to the field in mT measured by the sensor when there is no current. The slope has units of mT/turn and measures the rate at which the magnetic field increases as you add turns to the coil. The line should go through the origin, because the sensor was set to zero.

3. There is a linear relationship between magnetic field strength and current.

4. Equations should have the form $y = Ax + B$. The y-intercept corresponds to the field in mT measured by the sensor when there is no current. The slope is the field strength per ampere of current in mT/A. Because the sensor was set to zero, the line should go through the origin.

DATA TABLES WITH SAMPLE DATA

Part I

> **Observations – position and orientation of sensor**
> Answers will vary. The magnitude of the field will increase as the white dot turns toward magnetic north. The white dot on the sensor should point in the same direction as the needle of the magnetic compass.

Part II

Magnetic field versus turns parameters	
A (slope)	0.008 mT/turn
B (intercept)	−0.02 mT

Part III

Magnetic field versus current parameters	
A (slope)	0.26 mT/A
B (intercept)	0.011 mT

ANSWERS TO CONCLUSIONS

5. Adding more loops requires more wire and introduces more resistance into the circuit. More loops will also increase the strength of the magnetic field. Fewer loops decrease the strength of the field, but allow for less resistance in the circuit.

6. $current = \dfrac{(magnetic\ field\ strength - B)}{A}$

7. Answers will vary depending on the device tested.

8. Answers will vary depending on the device tested.

9. This meter would not have to be recalibrated for a 12 V system. The equation used to determine current from magnetic field strength is independent of voltage.

EXTENSIONS

1. There are many variations that can be done on this activity. One is to have a contest in which each group tries to produce the strongest magnetic field with a given length of wire. Students can also compete to design the smallest or most accurate current meter.

2. Students should find that the readings vary with Earth's magnetic field when they rotate the magnetic field sensor horizontally.

3. Students may describe the effects of solar sun spot activity on Earth's magnetosphere or any other current research into the variations of the magnetosphere.

TECHNOLOGY-BASED
LAB ACTIVITIES
CBL™ EXPERIMENTS

HOLT, RINEHART AND WINSTON

A Harcourt Classroom Education Company

Austin · New York · Orlando · Atlanta · San Francisco · Boston · Dallas · Toronto · London

Holt Physics

Technology-Based Lab Activities
CBL™ Experiments

Contributing Writer

E. David Thielk
Science Educator
Peninsula College
Port Townsend, WA

Reviewers

Richard Sorensen
Vernier Software & Technology
Beaverton, OR

Martin Taylor
Sargent-Welch/CENCO Physics
Buffalo Grove, IL

Safety Reviewer

Gregory Puskar
Laboratory Manager
Department of Physics
West Virginia University
Morgantown, WV

Cover photo: © Lawrence Manning/CORBIS
Cover design: Jason Wilson

Illustrations: All art is contributed by *Holt, Rinehart and Winston*.

Printed in the United States of America

ISBN 0-03-057342-4

2 3 4 5 6 095 05 04 03 02

Contents

Introduction to the Technology-Based Lab Activities

Modeling the physical world

Scientists and engineers often go to the laboratory to develop models of physical situations found in the world outside the lab. Each of the activities in this booklet presents a real-world scenario or problem for you to model in the laboratory. You will use your knowledge of physics to develop the model, and then you will test the model by collecting and analyzing data. After you have established a good working model, you will use the model to answer questions or solve problems in the real-world scenario.

Several different types of models may be used in a physics lab. A *physical model* is a physical representation of a real physical system, but the model is simplified and is often a small-scale representation. A *mathematical model* is a mathematical equation or set of equations that can be used to represent a physical phenomenon or relationship. A *graphical model* is a graph or set of graphs that can be used to represent the relationships between physical quantities. You will encounter all three of these types of models as you do the activities in this booklet.

As you work on these activities, you will develop a deeper understanding of how the concepts presented in your physics textbook relate to everyday physical phenomena, and you will use your understanding of physics to solve problems like those faced by physicists and engineers everyday.

Calculator-Based Laboratory™ (CBL™) technology

Calculator-Based Laboratory™ (CBL™) technology from Texas Instruments, coupled with software and probes by Vernier Software & Technology, turns your graphing calculator into a powerful lab instrument. The CBL is a separate unit that links to your graphing calculator with a black link cable. Various probes, such as motion detectors, force sensors, temperature probes, light sensors, or voltage probes, can be plugged into the CBL and allow your calculator and the CBL to collect physical data with a high degree of precision. You can then use your calculator to analyze the data, plot graphs, and calculate final results.

See page vii for information on downloading the necessary software for your graphing calculator. Each activity contains detailed instructions on using the CBL and probes. However, if you need additional information, ask your teacher about contacting TI or Vernier Software & Technology for CBL tutorials or technical support.

What you should do before a Technology Lab

Preparation will help you work safely and efficiently. Before each activity, be sure to do the following.

- **Download the PHYSICS application** if it is not already loaded onto your graphing calculator. See page vii for instructions.

- **Read the opening scenario** to understand the problem or situation you will be modeling in the lab.

- **Answer the questions** in the Developing the Model section. Your teacher may ask you to turn in your answers before doing the activity or as part of your final lab report.

- **Read the safety information** that begins on page viii, as well as the special safety instructions provided in each lab. Plan to wear appropriate clothing, shoes, and protective safety equipment while you work in the lab.

- **Read the procedure** to make sure you understand what you will do in each step.

- **Write down any questions** you have in your lab notebook so that you can ask them before the lab begins.

- **Prepare all necessary data tables** so that you will be able to concentrate on your work when you are in the lab.

What you should do after a Technology Lab

Most teachers require a written lab report as a way of making sure that you understood what you did in the lab. Your teacher will give you specific details about how to organize your written work for the Technology Labs, but most lab reports will include the following:

- **the title** of the activity

- **answers** to items and questions that appear in the Developing the Model section

- **data tables and observations** that are organized, complete, and easy to understand

- **answers** to items and questions that appear in the Analysis and Conclusions sections.

Each activity also includes extension exercises at the end. Your teacher may assign these or you may choose to do them on your own to further your understanding of the physical theories and real-world situations modeled in the lab.

Downloading Graphing Calculator Programs

---

internet connect

GO TO: go.hrw.com
KEYWORD: HF2 CALC
This Web site contains links for downloading programs and applications you will need for the Technology-Based Lab Activities.

To download required software for your graphing calculator, you will need a computer with an Internet connection, a TI-GRAPH LINK cable, and a TI-83 Plus or compatible calculator. Visit the go.hrw.com Web site, and type "HF2 CALC" at the keyword prompt.

1. If your computer does not already have TI-GRAPH LINK software installed, click Install TI-GRAPH LINK and follow the links for downloading and installing TI-Graph Link from the Texas Instruments Web site.

2. Click Download the PHYSICS App and follow the links for downloading the PHYSICS application from the Vernier Software & Technology Web site. You will also need to follow the instructions for your TI-GRAPH LINK to load the application onto your calculator. Once the PHYSICS application is loaded onto your calculator, it should appear in the APPS menu.

3. If you need more instructions on using the CBL system, click CBL Made Easy to view a tutorial from Vernier Software & Technology.

Note: The PHYSICS application can also be transferred directly between calculators using a unit-to-unit cable. Refer to the TI Web site or your calculator's user's manual for instructions.

Troubleshooting

• Calculator and CBL instructions in the *Holt Physics* program are written for the TI-83 Plus, for the original CBL from TI, and for probes from Vernier Software. If you use other hardware, some of the programs and instructions may not work exactly as described.

• If you have problems loading programs or applications onto your calculator, you may need to clear programs or other data from your calculator's memory.

• Always make sure that you are downloading correct versions of the software. TI-GRAPH LINK and the PHYSICS application both have different versions for different types of computers as well as different versions for different calculators.

• If you need additional help, both TI and Vernier Software can provide technical support.

HOLT PHYSICS Safety in the Physics Laboratory

Lab work is essential for progress in science and technology. Therefore, careful, systematic lab work is an essential part of any science program. In this class, you will practice some of the same fundamental laboratory procedures and techniques that experimental physicists use to pursue new knowledge.

The equipment and apparatus you will use involve various safety hazards, just as they do for working physicists. You must be aware of these hazards. Your teacher will guide you in properly using the equipment while doing the activities, but you must also take responsibility for your part in this process. With the active involvement of you and your teacher, these risks can be minimized so that working in the physics laboratory can be a safe, enjoyable process of discovery.

These safety rules always apply in the lab.

1. **Always wear a lab apron and safety goggles.**
   Wear these safety devices whenever you are in the lab, not just when you are working on an experiment.

2. **No contact lenses in the lab.**
   Contact lenses should not be worn during any investigations using chemicals (even if you are wearing goggles). In the event of an accident, chemicals can get behind contact lenses and cause serious damage before the lenses can be removed. If your doctor requires that you wear contact lenses instead of glasses, you should wear eye-cup safety goggles in the lab. Ask your doctor or your teacher how to use this important eye protection.

3. **Personal apparel should be appropriate for laboratory work.**
   On lab days, avoid wearing long necklaces, dangling bracelets, bulky jewelry, and bulky or loose-fitting clothing. Long hair should be tied back. Loose, dangling items may get caught in moving parts, accidentally contact electrical connections or interfere with the activity in some potentially hazardous manner. In addition, chemical fumes may react with some jewelry, such as pearls, and ruin them. Cotton clothing is preferable to wool, nylon or polyester. Wear shoes that will protect your feet from chemical spills and falling objects—no open-toed shoes or sandals, and no shoes with woven leather straps.

4. **NEVER work alone in the laboratory.**
   Work in the lab only while under the supervision of your teacher. Do not leave equipment unattended while it is in operation.

5. **Only books and notebooks needed for the activity should be in the lab.**
   Only the lab notebook and perhaps the textbook should be used. Keep other books, backpacks, purses and similar items in your desk, locker or designated storage area.

6. **Read through the entire activity before entering the lab.**
   Your teacher will review any applicable safety precautions before the activity. If you are not sure of something, ask your teacher about it.

7. **Always heed safety symbols and cautions written in the activities and handouts, posted in the room, and given verbally by your teacher.**
   They are provided for your safety.

8. **Know the proper fire drill procedures and location of fire exits and emergency equipment.**
   Make sure you know the procedures to follow in case of a fire or emergency.

9. **If your clothing catches on fire, do not run; WALK to the safety shower, stand under it, and turn it on.**
   Call to your teacher while you do this.

10. **Report all accidents to the teacher IMMEDIATELY, no matter how minor**.
    In addition, if you get a headache, feel sick to your stomach, or feel dizzy, tell your teacher immediately.

11. **Report all spills to your teacher immediately.**
    Call your teacher rather than trying to clean a spill yourself. Your teacher will tell you if you can safely clean up the spill; if not, your teacher will know how the spill should be cleaned up safely.

12. **Student-designed activities and experiments, such as those developed in extension exercises, must be approved by the teacher before being attempted by the student.**

13. **DO NOT perform unauthorized experiments or use materials and equipment in a manner for which they were not intended.**

    Use only materials and equipment listed in the activity equipment list or authorized by your teacher. Steps in a procedure should only be performed as described in the textbook or lab manual or as approved by your teacher.

14. **Stay alert in lab, and proceed with caution.**

    Be aware of others near you or your equipment when you are about to do something. If you are not sure of how to proceed, ask your teacher.

15. **Horseplay in the lab is very dangerous.**

    Laboratory equipment and apparatus are not toys. Never play in the lab or use lab time or equipment for anything other than their intended purpose.

16. **Food, beverages, chewing gum, and tobacco products are NEVER permitted in the laboratory.**

17. **NEVER taste chemicals. Do not touch chemicals or allow them to contact areas of bare skin.**

18. **Use extreme CAUTION when working with hot plates or other heating devices.**

    Keep your head, hands, hair and clothing away from the flame or heating area, and turn the devices off when they are not in use. Remember that metal surfaces connected to the heated area will become hot by conduction. Gas burners should be lit only with a spark lighter. Make sure all heating devices and gas valves are turned off before leaving the laboratory. Never leave a hot plate or other heating device unattended when it is in use. Remember that many metal, ceramic, and glass items do not always look hot when they are hot. Allow all items to cool before storing them.

19. **Exercise caution when working with electrical equipment.**

    Do not use electrical equipment with frayed or twisted wires. Be sure your hands are dry before using electrical equipment. Do not let electrical cords dangle from work stations; dangling cords can cause electrical shocks and other injuries.

20. **Keep work areas and apparatus clean and neat.**

    Always clean up any clutter made during the course of lab work, rearrange apparatus in an orderly manner, and report any damaged or missing items.

21. **Always thoroughly wash your hands with soap and water at the end of each activity.**

Safety Symbols

The following safety symbols will appear in the laboratory experiments to emphasize important additional areas of caution. Learn what they represent so you can take the appropriate precautions. Remember that the safety symbols represent hazards that apply to a specific activity, but the numbered rules given on the previous pages always apply to all laboratory work.

 Waste Disposal

- Never put broken glass or ceramics in a regular waste container. Use a dustpan, a brush, and heavy gloves to carefully pick up broken pieces and dispose of them in a container specifically provided for this purpose.

- Dispose of chemicals as instructed by your teacher. Never pour hazardous chemicals into a regular waste container. Never pour radioactive materials down the drain.

Heating Safety

- When using a burner or hot plate, always wear goggles and an apron to protect your eyes and clothing. Tie back long hair, secure loose clothing, and remove loose jewelry.

- Never leave a hot plate unattended while it is in use.
- Wire coils may heat up rapidly during this experiment. If heating occurs, open the switch immediately and handle the equipment with a hot mitt.
- Allow all equipment to cool before storing it.
- If your clothing catches on fire, walk to the emergency lab shower and use the shower to put out the fire.

 Hand Safety

- Perform this experiment in a clear area. Attach masses securely. Falling, dropped, or swinging objects can cause serious injury.
- Use a hot mitt to handle resistors, light sources, and other equipment that may be hot. Allow all equipment to cool before storing it.

 Glassware Safety

- If a thermometer breaks, notify the teacher **immediately.**
- Do not heat glassware that is broken, chipped, or cracked. Use tongs or a hot mitt to handle heated glassware and other equipment because it does not always look hot when it is hot. Allow all equipment to cool before storing it.
- If a bulb breaks, notify your teacher immediately. Do not remove broken bulbs from sockets.

Electrical Safety

- Never close a circuit until it has been approved by your teacher. Never rewire or adjust any element of a closed circuit.

- Never work with electricity near water; be sure the floor and all work surfaces are dry.
- If the pointer on any kind of meter moves off scale, open the circuit immediately by opening the switch.
- Do not work with any batteries, electrical devices, or magnets other than those provided by your teacher.

 Chemical Safety

- Do not eat or drink anything in the laboratory. Never taste chemicals or touch them with your bare hands.
- Do not allow radioactive materials to come into contact with your skin, hair, clothing, or personal belongings. Although the materials used in this lab are not hazardous when used properly, radioactive materials can cause serious illness when used improperly.

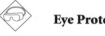 **Clothing Protection**

- Tie back long hair, secure loose clothing, and remove loose jewelry. Loose, dangling items may get caught in moving or rotating parts or come into contact with hazardous chemicals.

 **Eye Protection**

- Wear eye protection, and perform this experiment in a clear area. Swinging objects can cause serious injury.
- Avoid looking directly at a light source. Looking directly at a light source may cause permanent eye damage.

Chapter 1

HOLT PHYSICS
Technology Lab

Graph Matching

Testing Instructions for a Robotic Arm

Robotics, Inc., develops automated robotic arms used in manufacturing automobiles. The arms are designed to perform repetitive tasks quickly and consistently. A typical robotic arm consists of a tool that is able to pick up and drop objects, turn screws, pound in fasteners, or perform other mechanical tasks. The tool is mounted on a horizontal shaft that allows the tool to move back and forth to any position on the shaft. For example, a single robotic arm may be programmed to pick up a valve cover gasket, move horizontally toward an assembly line, place the gasket in position on the head of an engine, and then move back over to pick up another gasket. The next robotic arm on the assembly line may be programmed to pick up a valve cover and place it in position over the gasket.

Each robotic arm receives two sets of instructions. One set of instructions tells the tool attachment what it should be doing at a particular point in time. The other set of instructions tells the robotic arm where it should be at every point in time. The position instructions for the motion of a robotic arm can take the form of graphs that plot distance versus time or graphs that plot velocity versus time. In this activity, you will explore these two types of graphs as you model the motion of a robotic arm.

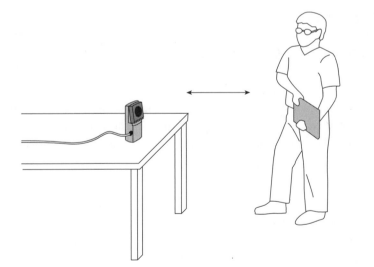

OBJECTIVES

- **Model** the motion of a robotic arm by moving an object in front of a motion detector and generating a distance-time graph.
- **Predict, sketch, and test** distance-time graphs that may be used as programming instructions for a robotic arm.
- **Predict, sketch, and test** velocity-time graphs that may be used as programming instructions for a robotic arm.
- **Evaluate** the use of graphs to provide instructions or analysis tools for the operation of robotic arms.

MATERIALS

- ✔ graphing calculator with link cable
- ✔ CBL system
- ✔ PHYSICS application loaded in calculator
- ✔ Vernier motion detector
- ✔ 2 ft. x 2 ft. piece of cardboard
- ✔ meterstick
- ✔ masking tape

SAFETY

- Review lab safety guidelines at the front of this booklet. Always follow correct procedures in the lab.
- Perform this experiment in a clear area. Be careful when moving around the room; other students may be in the process of gathering data with the motion detector.

DEVELOPING THE MODEL

A plot of the proper position of a robotic arm for a specific task can be generated by manually moving a robotic arm through the required task while the arm is monitored with a motion detector. The motion detector can collect data on the position of the arm at small time intervals, and a graphing calculator can then be used to present the data in graphic form. Once a graph is generated and stored, it can serve as a program—a set of instructions that can be repeated over and over again.

Before starting this activity, do the following exercises:

1. A robotic arm may have to move back and forth at different velocities and into different positions to complete a single task. To familiarize yourself with how this kind of information might look when stored as a graph, sketch a graph of distance versus time for each of the following situations:
   a) The robotic arm is at rest.
   b) The robotic arm is moving away from the reference point at a constant speed.
   c) The robotic arm is moving toward the reference point at a constant speed.
   d) The robotic arm is accelerating away from the reference point, starting from rest.

2. When you program a robotic arm, information about the speed at which the arm moves is important. To familiarize yourself with how this kind of information might look when stored as a graph, sketch a graph of velocity versus time for each of the situations described above.

PROCEDURE

Part I Generating Graphs

1. Lay out an area that represents the positions over which the robotic arm can move. Place the motion detector near the edge of a table, and point it toward an open space at least 3 m long. Secure the motion detector with a bracket or clamp. Use short strips of masking tape on the floor to mark the origin and distances of 1 m, 2 m, and 3 m from the motion detector.

2. Connect the Vernier motion detector to the SONIC port of the CBL unit. Use the black link cable to connect the CBL unit to the calculator. Firmly press in the cable ends.

3. Set up the calculator and CBL for the motion detector.
   • Start the PHYSICS application and proceed to the MAIN MENU.
   • Select SET UP PROBES from the MAIN MENU.
   • Select ONE as the number of probes.
   • Select MOTION from the SELECT PROBES menu.

4. Set up the calculator and CBL for data collection.
   • Select COLLECT DATA from the MAIN MENU.
   • Select TIME GRAPH from the DATA COLLECTION menu.
   • Enter "0.1" as the time, in seconds, between samples.
   • Enter "99" as the number of samples. The CBL will collect data for about 10 seconds.

- Press ENTER, then select USE TIME SETUP to continue. If you want to change the sample time or sample number, select MODIFY SETUP instead.
- Select LIVE DISP from the TIME GRAPH menu.
- Enter "0" for the Ymin so that the *y*-axis will start at 0 m.
- Enter "2.5" for the Ymax so that the *y*-axis will end at 2.5 m.
- Enter "0.5" for Yscl so that the *y*-axis will have a tick mark every half meter.

5. Now you will simulate the motion of a robotic arm and make a graph of the motion. Hold the piece of cardboard in front of the motion detector, and have your lab partner press ENTER. When you hear the motion detector begin to click, slowly move the piece of cardboard away from the motion detector. Be careful not to trip or bump into anyone or anything as you move. As you move the cardboard, a graph will be generated on the graphing calculator. Stop moving when the motion detector stops clicking. After data collection is complete, press ENTER to see a rescaled version of the graph.

6. Practice step 5 until you get a fairly smooth, straight line on the graph. To collect more data, press ENTER and select YES from the REPEAT? menu. Sketch the best final graph that you have produced.

7. Try to match the shape of each of the four distance-time graphs that you sketched in the first exercise in the Developing the Model section.

8. Perform steps 1–3 of the Analysis (on page 5) before proceeding to Part II of the Procedure.

Part II Distance-Time Graph Matching

9. Return to the MAIN MENU by selecting NO from the REPEAT? menu.

10. Now you will test the use of distance-time graphs to convey information about the movement of a robotic arm. The PHYSICS application can generate random distance programs for the robotic arm, such as the sample shown below. The graph your calculator generates will probably have a different curve.
- Select COLLECT DATA from the MAIN MENU.
- Select GRAPH MATCH from the DATA COLLECTION menu.
- Select DISTANCE MATCH from the GRAPH MATCH menu.
- Note the screen instructions, and press ENTER.

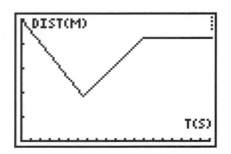

11. Sketch a copy of the graph that appears on your calculator screen. The vertical axis (distance) runs from 0 to 2.5 m, and the horizontal axis (time) runs from 0 to 10 s. Write down how the robotic arm would have to move to produce this target graph.

12. To test your prediction, choose a starting position and hold the cardboard at that point. Start data collection by pressing ⌽ENTER. When you hear the motion detector begin to click, move the cardboard in such a way that the graph of the cardboard's motion matches the target graph on the calculator screen. Be careful not to trip or bump into anyone or anything as you move.

13. If you were not successful, repeat step 12 until the graph of the cardboard's motion closely matches the graph on the screen. To repeat with the same graph, press ⌽ENTER and select SAME MATCH from the OPTIONS menu. Sketch the graph with your best attempt.

14. Perform a second distance graph match (steps 10–12) by pressing ⌽ENTER and selecting NEW MATCH from the OPTIONS menu.

15. Perform steps 4–8 of the Analysis (on page 5) before proceeding to Part III of the Procedure.

Part III Velocity-Time Graph Matching

16. Now you will test the use of velocity-time graphs to convey information about the movement of a robotic arm. The PHYSICS application can generate velocity-time graphs similar to the one shown below. The graph your calculator generates will probably have a different curve.

- Select RETURN TO MAIN from the OPTIONS menu.
- Select COLLECT DATA from the MAIN MENU.
- Select GRAPH MATCH from the DATA COLLECTION menu.
- Select VELOCITY MATCH from the GRAPH MATCH menu.
- Note the screen instructions, and press ⌽ENTER.

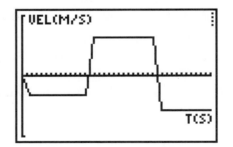

17. Sketch a copy of the graph on your calculator screen. The vertical axis (velocity) runs from –0.5 to +0.5 m/s, and the horizontal axis (time) runs from 0 to 10 s. Write down how the robotic arm would have to move to produce this target graph.

18. To test your prediction, choose a starting position and hold the cardboard at that point. Start data collection by pressing ⌽ENTER. When you hear the motion detector begin to click, move the cardboard so that the graph of the cardboard's motion matches the target graph on the calculator screen. Be careful not to trip or bump into anyone or anything as you move.

19. If you were not successful, repeat step 18 until the motion of the cardboard closely matches the graph on the screen. To repeat with the same graph, press ⌽ENTER and select SAME MATCH. Sketch the graph with your best attempt.

20. Perform a second velocity graph match (steps 16–18) by pressing ⌽ENTER and selecting NEW MATCH from the OPTIONS menu.

21. Remove the masking tape strips from the floor. Clean up your work area. Put equipment away safely so that it is ready to be used again. Recycle or dispose of used materials as directed by your teacher.

22. Perform steps 9–11 of the Analysis (below), then continue to the Conclusions.

ANALYSIS

Part I

1. Communicating results Describe the shape of the distance-time graph you generated in step 5 of the Procedure.

2. Making predictions How would you have to move to produce a straight line sloping down to the right on a distance-time graph?

3. Interpreting graphs Were the graphs on the calculator screen always smooth and even, or did they sometimes have spikes and other discontinuities? If the graphs had spikes, why do you think the spikes appeared?

Return to the Procedure and complete Part II (steps 9–15).

Part II

4. Communicating results Describe how the position of the cardboard changed for each of the distance-time graphs that you matched.

5. Interpreting graphs Each distance-time graph may have several different slopes corresponding to different aspects of the motion. What does a positive or negative slope indicate about the motion of the robotic arm?

6. Interpreting graphs What type of motion is occurring when the slope of a distance-time graph is zero?

7. Interpreting graphs What does the magnitude of the slope indicate about the motion?

8. Interpreting graphs What type of motion is occurring when the slope of a distance-time graph is changing? Use the motion detector to test your answer to this question.

Return to the Procedure and complete Part III (steps 16–22).

Part III

9. Communicating results Describe how the speed of the cardboard changed over time for each of the velocity-time graphs that you matched.

10. Interpreting graphs What type of motion is occurring when the slope of a velocity-time graph is zero?

11. Interpreting graphs What type of motion is occurring when the slope of a velocity-time graph is not zero? Use the motion detector to test your answer.

CONCLUSIONS

12. Evaluating models In this activity, you interpreted data in two different forms: distance versus time and velocity versus time. Which form was the easiest to interpret? Why do you think so?

13. Applying results If a motion detector were used to monitor the position of a robotic arm, which of these two types of graphs would be most useful to production engineers? Explain why you think so.

14. **Applying theory** Imagine that the programming instructions are in the form of distance-time graphs, and the production manager wants to speed production by 10 percent. How would the graphs have to be changed to reflect a 10 percent increase in production?

15. **Applying theory** Imagine that the programming instructions were in the form of velocity-time graphs and the production manager wanted to speed production by 10 percent. How would the graphs have to be changed to reflect a 10 percent increase in production?

EXTENSIONS

1. **Making predictions** Devise a task that would require a robotic arm to move back and forth horizontally. Make sure the required path length is short enough to fit in the space you have available. Ask your lab partner to sketch a distance-time graph for the task and to then use the motion detector to generate a distance-time graph for the task on the calculator.

2. **Analyzing systems** Create a simple pendulum by hanging a lab balance mass or other object from a string that is about 1 m long. Hold the pendulum in your hand, and gently set it in motion using occillations less than 15° from the rest position. Observe the back-and-forth motion of the pendulum, then sketch a distance-time graph and a velocity-time graph for the pendulum's motion. Use the motion detector and the CBL to graph the motion of the pendulum, and compare the graphs to the sketches you created.

HOLT PHYSICS
Technology Lab A

Acceleration

Measuring G-forces in a Bungee Jump

You have just been hired by Outdoor Adventures as head of the Physics of Thrill Department. The company has informed you that several companies in New Zealand have switched from a traditional bungee cord (using a latex core surrounded by a woven sheath) to cords made of pure latex. These companies are advertising that their new cords allow clients to make longer, higher bungee jumps with less stress on the body. To remain competitive and to be able to continue to attract clients, Outdoor Adventures would like to develop its own latex cord.

Your first task as head of the department is to develop a laboratory model that can be used to test different grades and configurations of latex cords. The goal of each test is to measure the effect of the jump on the body. The primary quantity you will measure is the acceleration that the bungee jumper's body would experience during the jump.

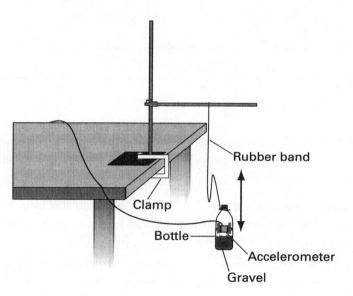

OBJECTIVES

- **Model** the motion of a bungee jumper, and use an accelerometer to measure accelerations.
- **Compare** the data from the model with an actual bungee jump.
- **Propose** suggestions to improve the model.

MATERIALS

- ✔ graphing calculator with link cable
- ✔ CBL system
- ✔ PHYSICS application loaded in calculator
- ✔ Vernier low-g accelerometer with CBL adapter cable
- ✔ table clamp with rod
- ✔ small plastic bottle with gravel inside
- ✔ 0.5 m stretchable cord (bungee cord, a long rubber band or several rubber bands tied together, latex, waistband elastic, surgical tubing, etc.)
- ✔ meterstick

SAFETY

- Perform this experiment in a clear area. Attach masses securely. Falling, dropped, or swinging masses can cause serious injury.
- Tie back long hair, secure loose clothing, and remove loose jewelry to keep them from getting caught in moving or rotating parts.

DEVELOPING THE MODEL

Large rates of acceleration experienced by the human body are often measured in units of the free-fall acceleration g. The body of a healthy person can safely withstand accelerations of about five times the free-fall acceleration, or $5g$, for periods of a few seconds.

As a guideline, you have been given a graph of acceleration versus time for a real bungee jumper using a standard bungee cord (below). In real bungee jumping, the bungee cord should stretch from 2.5 to 4 times its original length before stopping. In addition, the bungee cord should be matched to the weight of the jumper so that it still has a significant ability to stretch when the jumper reaches the bottom of the ride. The jumper should also experience at least two bounces with little or no tension in the bungee cord. Your laboratory model should approximate these specifications.

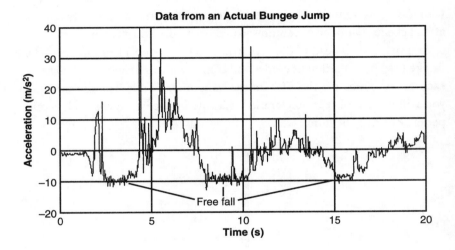

When setting up your model, start by checking the following characteristics of the bungee cord you will be using. Write answers to the questions before going on to the Procedure section.

1. Determine the cord's ability to stretch. You can do this by measuring the length of a small section, then stretching it and measuring the stretched length. Describe in writing the ability of your cord to stretch.

2. You will use a plastic bottle filled with gravel to model the jumper. Securely attach the cord with a solid knot through an eyebolt in the lid or around the neck of the bottle. Ask for help if necessary. Hold the end of the cord in one hand and the bottle in the other hand. Make sure the bottle will not hit your feet or any other students, then drop the bottle. Use the meterstick to determine if the bottle causes the cord to stretch to at least two times its original length. Make sure that the cord does not stretch to its maximum length and that the bottle does not hit the ground. Add gravel to or remove gravel from the bottle to adjust the model, always making sure you securely tighten the lid. Comment on the range of displacement experienced by the jumper.

3. When you dropped the jumper, did the jumper experience two bounces in which the cord lost all or nearly all of the tension? Explain how you could tell when this happened.

PROCEDURE

1. Connect the accelerometer to CH 1 of the CBL unit. Use the black link cable to connect the CBL unit to the calculator. Firmly press in the cable ends. Use tape to securely attach the accelerometer to the plastic bottle. The arrow on the accelerometer should point upward.

2. Securely attach the cord to the neck of the bottle or to an eyebolt in the lid of the bottle. Securely attach the other end of the cord to a rod that is clamped to the edge of the lab table. Adjust the length of the cord so that the bottle does not hit the floor when dropped. Ask your teacher for help if necessary. Measure and record the unstretched length of the cord.

3. Turn on the CBL unit and the calculator. Start the PHYSICS application and proceed to the MAIN MENU.

4. Set up the calculator and CBL for the accelerometer.
   - Select SET UP PROBES from the MAIN MENU.
   - Select ONE as the number of probes.
   - Select ACCELEROMETER from the SELECT PROBE menu.
   - Confirm that the accelerometer is connected to CHANNEL 1, and press [ENTER].
   - Select USE STORED from the CALIBRATION menu.
   - Select LOW-G from the ACCELEROMETER menu.

5. The accelerometer must be calibrated so that it records for the vertical direction only. The accelerometer should record zero acceleration when at rest and -9.8 m/s^2 during free fall. You will verify this later in steps 8–9.
   - Rest the bottle on the table, so that the accelerometer arrow points directly upward.
   - Select ZERO PROBES from the MAIN MENU.
   - Select CHANNEL 1 from the SELECT CHANNEL menu.
   - When the reading on the screen is stable, follow the instructions on the calculator to zero the sensor.

6. Set up the calculator and CBL for data collection.
   - Select COLLECT DATA from the MAIN MENU.
   - Select TIME GRAPH from the DATA COLLECTION menu.
   - Enter "0.05" as the time between samples in seconds.
   - Enter "80" as the number of samples.
   - Press [ENTER], and select USE TIME SETUP to continue. To change the sample time or sample number, select MODIFY SETUP instead.

7. Hold the bottle at the height of the rod. Make sure the arrow on the accelerometer points up. Press [ENTER] to begin collecting data. Hold the bottle motionless for one second, then release it. Catch the bottle before it reaches the bottom of its fall, while the cord is still slack.

8. When data collection has finished, press [ENTER] to see your graph. Trace across the graph with the cursor keys. For the first second or so, the acceleration should be near zero. This value represents the acceleration of the bottle before it began to fall.

9. Trace farther to the right on your graph and read the acceleration during the fall. It should be close to -9.8 m/s^2. Ignore any data points collected after you caught the bottle. If the result is not close to -9.8 m/s^2, press ENTER and select YES to prepare to collect more data, then repeat steps 7–9.

10. Next, you will collect data that corresponds to the bounces after the free-fall portion of the jump. Make sure the area directly beneath the horizontal rod and bottle is clear.

   • Let the bottle hang from the cord.

   • Press ENTER and select YES to prepare to collect more data.

   • Pull the bottle down 5 cm and hold it stationary.

   • Press ENTER to start data collection.

   • Release the bottle. This will cause the bottle to oscillate up-and-down like a mass suspended from a vibrating spring.

   • When the data collection is finished, press ENTER to view the graph. Find the point in the motion where acceleration is positive in direction and has a maximum magnitude. Does this occur when the bottle is at the bottom, middle, or top of the oscillation?

11. Lift the bottle to the height of the horizontal rod. The cord should be hanging to the side, and the accelerometer cable should be clear of the jump path. Make sure the arrow on the accelerometer points up. The bottle should also be pointing upward so that it will not turn over during the fall.

   • Press ENTER and select YES to prepare to collect more data.

   • Press ENTER to start collecting data.

   • Wait one second, then release the bungee jumper so that it falls straight down with a minimum of rotation. Let the bottle bounce a few times. Make sure that the accelerometer cable still has some slack when the bottle reaches its lowest point.

12. Repeat step 11 until you have a set of data that includes a minimum of rotation, a section of free fall before the cord starts to pull on the bottle, and a few bounces, with at least the first bounce high enough to allow the cord to again go slack. Sketch your final graph.

DATA TABLE

Length of cord: _____ m

Time (s)	Acceleration (m/s²)	Direction of motion (up, down, or at rest)

ANALYSIS

1. **Graph tracing** Use the cursor keys to trace across the graph of acceleration versus time. Determine the acceleration at eight different points on the graph, choosing points during the initial rest, free fall, when the cord is taut, and several bounces. Record the time and acceleration values in the data table. Indicate the direction of the motion at each point using *up, down*, or *at rest*.

2. **Interpreting graphs** On your graph, find and label the first point at which the bungee cord had no slack. Also find and label this point on the graph of the data from the actual bungee jump. Explain how you can tell when this moment occurs.

3. **Interpreting graphs** On your graph and on the graph of the actual bungee jump data, find and label the moment at which the bungee cord was stretched to a maximum. Explain how you can tell when this moment occurs.

4. **Analyzing data** At the moment of maximum stretch, what was the acceleration of the bottle? What was the acceleration of the jumper at the corresponding point in the actual bungee jump? Was the acceleration at this point upward or downward?

5. **Analyzing data** Compare the acceleration of the bottle at the moment of maximum stretch and the velocity of the bottle at the same moment. What is the relationship between the two quantities?

6. **Interpreting graphs** Find and label the moment during the first bounce in which the bottle was at the highest point. Explain how you can tell when this point occurs.

CONCLUSIONS

7. **Evaluating models** Compare the similarities and differences between the model and the real bungee jump data. Include a discussion of maximum acceleration, the duration of the jump, and the number of bounces. Is your model useful? Explain.

8. **Making predictions** Predict what the acceleration-time graph would look like if you changed some of the variables in the procedure.

   a. How would using a heavier jumper affect the model?

   b. How would changing the length of cord affect the model?

   c. How would using a cord that required more force to stretch it affect the model? Explain.

EXTENSIONS

1. **Applying results** Calculate the length of the bungee cord used in the real bungee jump data. Remember that the distance an object falls under constant acceleration is given by the equation $d = \frac{1}{2}gt^2$. Perform the same calculation on the data you collected and compare the result against the length of the cord you used.

2. **Comparing results** Repeat the experiment with a bottle of different mass. Make sure the cord you use is appropriate, so that it stretches substantially but does not reach its maximum extension or break. What are the similarities and differences between the two sets of data? Discuss some methods that might be used by operators of commercial bungee jumps to assure the safety of jumpers of different weights.

3. **Extending research** Use reference books or the Internet to research the accelerations experienced by astronauts during takeoff and re-entry. How do the accelerations experienced by astronauts compare to the maximum acceleration experienced by a bungee jumper?

4. **Evaluating methods** Place a motion detector on the floor under the bottle during a "jump." To protect the motion detector, place a wire basket upside-down over the detector and make sure the bottle will not hit the basket when it falls. Examine the distance versus time and velocity versus time graphs of the jump from the motion detector data. How do these data compare with the accelerometer data? Which sensor do you think is a better tool for analyzing a jump? Explain.

2 Technology Lab B

Free Fall

Galileo's Assistant and the Case of the Variable *g*

You have just taken on a summer job as an assistant in a small private research lab in Pisa, Italy. Your boss, Galileo, has determined through a series of careful experiments that objects of different masses roll down an inclined plane at the same rate. He has further hypothesized, based on these experiments, that objects in free fall should always fall at the same rate, no matter what their masses are. This result contradicts the popular theory of the ancient Greek philosopher Aristotle (384–322 BC), who believed that all falling objects travel downward at speeds directly proportional to their masses.

In one preliminary experiment, Galileo drops a wooden ball and a crumpled piece of parchment at the same time from near the ceiling of the lab. Instead of reaching the ground at the same time, the wooden ball lands shortly before the parchment does, apparently supporting Aristotle's theory. Galileo suspects that other factors are coming into play, such as air resistance. However, he also knows that opponents of his theory would claim that the two objects experience different free-fall accelerations because they have different masses.

Your assignment is to help Galileo by accurately measuring the free-fall acceleration, *g*, of several objects. If you find that all the objects fall with the same rate of acceleration, you will confirm Galileo's hypothesis. Short of that, you may disprove Aristotle's hypothesis if you find that a heavier object sometimes falls with the same acceleration as—or even with lesser acceleration than—a lighter object.

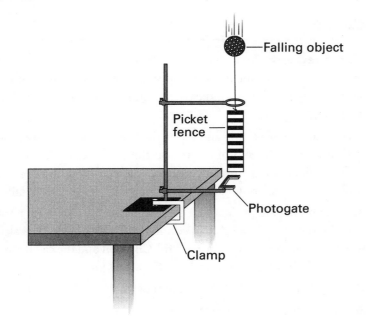

OBJECTIVES

- **Measure** the acceleration of several falling objects of different masses.

- **Compare** measured free-fall accelerations with an accepted standard value.

- **Determine** whether or not the free-fall acceleration of an object depends directly on the object's mass.

MATERIALS

✔ graphing calculator with link cable

✔ CBL system

✔ PHYSICS application loaded in calculator

✔ Vernier photogate with CBL adapter

✔ Vernier picket fence

✔ table clamp and rod

✔ support ring with clamp

✔ additional clamp (for attaching photogate)

✔ 1 m piece of string

✔ assorted objects: plastic ball, crumpled newspaper, baseball, balloon, etc.

✔ cushion, pillow, or padded catch box

✔ strong, clear tape

SAFETY

- Perform this experiment in a clear area. Attach masses securely. Falling, dropped, or swinging masses can cause serious injury.

- Tie back long hair, secure loose clothing, and remove loose jewelry to keep them from getting caught in moving or rotating parts.

DEVELOPING THE MODEL

Fortunately, you have access to rather sophisticated equipment to help Galileo. You will use a CBL unit connected to a photogate and a graphing calculator. You will measure acceleration using a "picket fence," a transparent plastic sheet with evenly spaced black bars. As the picket fence falls through the photogate, the bars of the picket fence will interrupt the infrared beam between the arms of the photogate. The CBL will measure the time from the leading edge of one bar blocking the beam until the leading edge of the next bar blocks the beam. This timing continues as all eight bars pass through the photogate. From these measured times, the PHYSICS application calculates the velocities for each of the intervals between the bars of the picket fence.

Before starting this activity, answer the following questions.

1. The CBL unit measures the time interval between each black bar on the picket fence. What other information would you need to determine the average velocity between any two of the black bars?

2. If the picket fence is not vertical when it passes through the photogate, how might that affect your data?

3. Sketch a graph of velocity versus time for an object in free fall. What does the shape of the graph suggest about the acceleration of the object as it falls?

4. What are the units of the slope of a line on a graph of velocity versus time? What physical quantity does this slope represent?

PROCEDURE

Part I Picket fence in free fall

1. Prepare a data table like the first one shown in the Data Tables section. Use a balance to measure the mass of the picket fence, and record the mass in your data table.

2. Securely anchor a table clamp and rod to the edge of a table. Fasten the photogate to the lower portion of the rod so the arms extend horizontally, as shown in the figure on the previous page. Place a cushion or a catch box on the ground under the photogate to catch the picket fence after it falls.

3. Connect the photogate to the CH 1 input on the CBL using the photogate adapter. Use the black link cable to connect the CBL unit to the calculator. Firmly press in the cable ends. Turn on the CBL unit and the calculator. Start the PHYSICS application and proceed to the MAIN MENU.

4. Set up the calculator and CBL for the photogate.

    - Select SET UP PROBES from the MAIN MENU.

    - Select ONE as the number of probes.

- Select PHOTOGATE from the SELECT PROBE menu.

- Proceed to the TIMING MODES menu.

- Select CHECK GATE to see that the photogate is functioning.

- Block the photogate with your hand; note on the screen that the photogate is shown as blocked. Remove your hand and the display should change to unblocked.

- Press ⎡ + ⎤ to return to the PHOTOGATE TIMING menu.

5. Set up the calculator and CBL for timing with the picket fence.

- Select MOTION from the TIMING MODES menu.

- Select SELECT DEVICE from the MOTION TIMING menu.

- Select VERNIER PICKET from the SELECT DEVICE menu.

6. Now you are prepared to collect free-fall data. Select COLLECT DATA from the MOTION TIMING menu, and press ⎡ENTER⎤ to start collecting data.

7. Hold the picket fence in a vertical position just above the photogate, grasping it by the top edge. Drop the picket fence through the photogate, releasing it from your grasp completely before it crosses the infrared photogate beam. The picket fence must not touch the photogate as it falls, but the black bars must interrupt the infrared beam. The picket fence should remain vertical as it passes through the photogate.

8. Select VELOCITY to display a graph of velocity versus time. Sketch the graph on paper for later use. **Proceed to step 1 of the Analysis before continuing.**

9. To establish the reliability of your data, repeat steps 6–8 four more times. Do not use drops in which the picket fence hits or misses the photogate or drops in which the photogate does not remain vertical while falling. Record the slope values for each trial in your data table.

Part II Objects of different masses in free fall

10. Now you will find the free-fall accelerations of several objects in free fall while attached to the picket fence. These objects (plastic ball, crumpled paper, etc.) will all be about the same size and shape, to minimize the dependence on other effects such as air resistance. Securely clamp a small support ring to the rod, above the photogate, leaving at least 40 cm between the ring and the photogate. This ring will keep the objects you drop from hitting the photogate. Use a piece of strong, clear tape to securely attach a piece of string to the picket fence.

11. Use a balance to measure the combined mass of the picket fence, the string, and the first object you will be using. Record the mass in your data table. If you already know the other objects you will be using, you may want to measure their masses (combined with the picket fence and string) now.

12. Run the string up through the support ring, and then securely tie or tape the other end of the string to the first object you will be using. Make sure the cushion or catch box is still directly under the picket fence and photogate.

13. Repeat steps 6–9 for a total of five trials, each time holding and releasing the object so that the picket fence falls through the photogate. Your lab partner may need to steady the picket fence to stop it from swinging before you release the object. After each trial, perform step 1 of the Analysis and record the slope values in your data table.

14. Repeat steps 11–13 three or four times with different objects attached to the picket fence. Record all slope values in your data table. When you are finished, proceed to step 2 of the Analysis.

DATA TABLES

Object	Mass (kg)	Slope (trial 1)	Slope (trial 2)	Slope (trial 3)	Slope (trial 4)	Slope (trial 5)
picket fence alone						
plastic ball						
crumpled paper						
baseball-sized balloon						
baseball						

Object	Average slope	Free-fall acceleration (m/s²)	Percent uncertainty
picket fence alone			
plastic ball			
crumpled paper			
baseball-sized balloon			
baseball			

ANALYSIS

1. **Curve fitting** The slope of a velocity-time graph is a measure of acceleration. If the velocity graph is approximately a straight line of constant slope, the acceleration is constant. You may fit a straight line to your data in the following way:

    - Press ⟨ENTER⟩, select NEXT from the SELECT GRAPH menu.
    - Select NO, and then RETURN TO MAIN from the TIMING MODES menu.
    - Select ANALYZE from the MAIN MENU.
    - Select CURVE FIT from the ANALYZE MENU.
    - Select LINEAR L₁, L₅ from the CURVE FIT menu.
    - Record the slope of the fitted line in your data table.
    - Press ⟨ENTER⟩ to see the fitted line superimposed on your velocity-time graph.

    If you have not yet completed the Procedure, return to step 9 in the Procedure now.

2. **Calculating averages** Prepare a data table like the second one shown in the Data Tables section. For each object, including the picket fence alone, use the slope values for all five trials to calculate the average slope value for that object. Record the average slopes in your new data table. These averages are

the rates of free-fall acceleration for each object. Note, however, that the PHYSICS application treats all velocities as positive, so the slopes are also positive. The free-fall acceleration, however, is negative, because the downward direction is negative by convention.

3. **Calculating uncertainty** The minimum and maximum values among the trials with each object give an indication of *uncertainty,* or how much the measurements can vary from trial to trial. One simple way to calculate uncertainty is to take half the difference between the minimum and maximum values over several trials, then round the uncertainty to one digit and round the average value to the same decimal place. Calculate uncertainty in this way for each object. Record your final answers as *−average ± uncertainty* under the heading *Free-fall acceleration* in your data table (the initial minus sign indicates that the acceleration is downward).

4. **Calculating percent uncertainty** Uncertainty is often expressed as a percentage. Calculate the percent uncertainty of your free-fall acceleration for each object using the following equation, and record your results in your data table.

$$percent\ uncertainty = \frac{uncertainty}{average} \times 100\%$$

CONCLUSIONS

5. **Determining accuracy** Compare your value of the free-fall acceleration for the picket fence alone to the generally accepted value of *g*, 9.81 m/s². Does the accepted value fall within the range of your values?

6. **Comparing values** Which objects have a free-fall acceleration closest to the free-fall acceleration of the picket fence alone? Which objects have free-fall accelerations that are farthest from the free-fall acceleration of the picket fence alone?

7. **Interpreting results** For the object with the free-fall acceleration farthest from that for the picket fence alone, does the free-fall acceleration, including uncertainty, lie completely outside the value of free-fall acceleration for the picket fence alone, including uncertainty? Does this support or contradict Galileo's hypothesis that free-fall acceleration is independent of mass?

8. **Making predictions** If Galileo dropped a baseball and a baseball-sized balloon from the Leaning Tower of Pisa, would he be able to demonstrate that all objects experience the same free-fall acceleration? Explain.

9. **Formulating hypotheses** If all the objects you tested did not fall with the same rate of acceleration, propose an explanation of the variations in your measured values of free-fall acceleration that you or Galileo could offer to critics of Galileo's hypothesis.

10. **Interpreting results** Do your results show, for any two objects, a heavier object accelerating at the same rate as a lighter object, or even accelerating at a slower rate than a lighter object? Does this support or contradict Aristotle's hypothesis that free-fall acceleration is directly proportional to mass?

EXTENSIONS

1. **Graphing** Plot a graph of free-fall acceleration versus mass for the various objects you tested. Above and below each data point, draw short lines that span the length of the uncertainty on either side of the average acceleration values. Try to find a line that intersects all of the data points. Does your result support or contradict Aristotle's hypothesis that free-fall acceleration is directly proportional to mass?

2. **Curve fitting** Galileo's study of free-fall acceleration led, among other things, to the following quadratic equation for displacement with constant acceleration:

$$\Delta x = v_i \Delta t + \frac{1}{2}a(\Delta t)^2$$

Run the experiment again with the picket fence alone, and this time display a graph of distance versus time. Perform step 1 of the Analysis again, but this time select QUADRATIC L_1, L_5 from the CURVE FIT menu. Sketch the resulting graph and fitted curve. What geometric shape is the fitted curve?

3. **Extending research** Research how the value of free-fall acceleration, *g*, varies at different locations around the world. For example, how does altitude affect the value of *g*? How much can *g* vary at a location in the mountains compared to a location at sea level? What other factors may cause free-fall acceleration to vary at different locations?

Technology Lab

Projectile Motion

Planning for A Car-Crash Stunt Scene

The director of a movie is planning a scene in which an expensive sports car will be driven off a vertical cliff onto the beach below. To achieve the most dramatic effect, she wishes to have the car crash onto the beach right at the water's edge. The cliff is 112.0 m high, and the targeted spot is between 38 m and 42 m from the base of the cliff.

Because the car will be ruined in the scene, the director wants the crash to occur as planned on the first take. She needs to know how fast the car must travel so that it will hit the beach in the correct spot. In this lab, you will model this scenario by rolling a ball off a table, and you will collect data on the motion of the ball to help you determine the proper speed of the car in the movie scene.

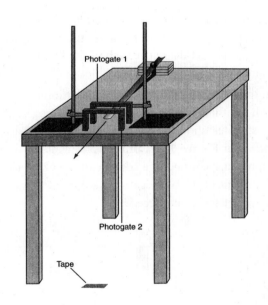

Photogate 1

Photogate 2

Tape

Figure 3-1

OBJECTIVES

- **Develop** a model that can be used to simulate a car driving off a cliff on a movie set.
- **Analyze** the relationship between the horizontal velocity and impact point of a projectile, using this model.
- **Evaluate** the assumptions made by the model.
- **Predict** the velocity required for the car to land at the targeted spot on the beach.

MATERIALS

- ✔ graphing calculator with link cable
- ✔ CBL system
- ✔ PHYSICS application loaded in calculator
- ✔ two Vernier photogates
- ✔ Vernier dual photogate adapter
- ✔ two support stands
- ✔ two right-angle clamps
- ✔ steel ball (1 to 3 cm diameter)
- ✔ ramp
- ✔ plumb bob
- ✔ meterstick or metric measuring tape
- ✔ piece of cardboard or catch box
- ✔ masking tape
- ✔ carbon paper

SAFETY

- Perform this experiment in a clear area. Objects rolling loose in the laboratory can be hazardous. Use a piece of cardboard or a catch box to stop rolling balls.

- Tie back long hair, secure loose clothing, and remove loose jewelry to prevent them from getting caught in moving or rotating parts.

DEVELOPING THE MODEL

As you model the car crash with the rolling ball, you will be measuring the horizontal velocity of the ball and the distance the ball travels before striking the floor. Before starting this lab, answer the following questions:

1. When the car in the movie leaves the cliff, what factors will affect the time the car takes to hit the beach? Compare this scenario to a ball rolling across a tabletop and falling to the floor. How is this similar? How is it different?

2. How will the horizontal velocity of the ball or the car affect the distance that it travels before striking the ground? Which of these quantities is the independent variable? Sketch a graph that describes the relationship between horizontal velocity and distance.

3. The speedometer can be used to accurately determine the velocity of the car. In the model, the ball will roll through a pair of photogates. The CBL will be used to measure the time interval between the ball breaking the beam of one photogate and then breaking the beam of the other photogate. How can this measurement be used to determine the velocity of the ball? What other measurements will you need?

PROCEDURE

1. Set up a ramp made of angle molding on a table so that a ball can roll down the ramp, across a short section of the table, and off the edge, as shown in **Figure 3-1**. Make sure that the entire path of the ball will be clear throughout the experiment.

2. Attach the photogates to the support stands. Position the support stands so that the ball will roll through the middle of each photogate after it leaves the ramp. The photogates should be placed approximately 3 cm from one another, as shown in **Figure 3-2**. Connect photogates 1 and 2 to the dual photogate adapter, and connect the adapter to the CH 1 input of the CBL. To prevent movement of the photogates, use clamps, masses, or strong tape to secure the support stands in place.

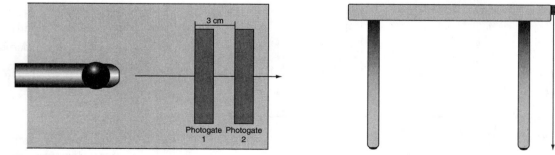

Figure 3-2

Figure 3-3

3. Mark three starting positions on the ramp. Practice rolling the ball down the ramp, through the photogates, and off the table. Make sure that the ball does not strike the sides of the photogates. Adjust the photogates if necessary. Use a catch box or a piece of cardboard folded in a "V" to prevent the ball from rolling across the room after the ball hits the floor.

4. Carefully measure the distance between the two photogates. The accuracy of the model depends on this measurement. Record the distance as d in a data table like the one in the Data Table section.

5. Carefully measure the distance from the top of the table to the floor, and record the measurement as the table height, Δy, in your data table. Use a plumb bob to locate the point on the floor directly beneath the point where the ball will leave the table, as shown in **Figure 3-3.** Mark this point with tape.

6. Turn on the CBL unit and the calculator. Start the PHYSICS application and proceed to the MAIN MENU.

7. Set up the calculator and CBL for the photogates.
   - Select SET UP PROBES from the MAIN MENU.
   - Select ONE as the number of probes. (Even though you are using two photogates, they use a single channel of the CBL. As a result, PHYSICS treats them as one probe.)
   - Select PHOTOGATE from the SELECT PROBE menu.
   - Press ENTER to proceed to the TIMING MODES menu.

8. Select CHECK GATE. Observe the reading on the calculator screen. Block each photogate in turn with your hand; note that the photogate is shown as blocked on the calculator screen. Remove your hand, and the display should change to unblocked. Press + to return to the TIMING MODES menu.

9. The calculator will measure the time interval between the blocking of the first gate and the blocking of the second gate. You can see how this works by blocking one gate and then the other. Prepare the calculator to do this.
   - Select PULSE from the TIMING MODES menu.
   - Press ENTER to arm the gates.
   - Block one gate with your hand, and then remove your hand from the gate.
   - Block the other gate.

   Note that the calculator now shows a time interval in seconds. This is the time interval between blocking the first gate and blocking the second gate.

10. To model the point of impact of the car as it hits the beach, you will need to measure the horizontal distance that the ball travels before hitting the floor. Roll the ball from the lowest mark on the ramp through the photogates, and watch where the ball hits the floor. Place your carbon paper over this spot with the ink side facing downward. Move your catch box or cardboard as necessary. Press ENTER and select YES to prepare the calculator for the next trial.

11. You are now ready to collect data. Make sure the path of the ball's motion is clear, then roll the ball from the lowest mark on the ramp. Locate the mark on the floor made by the carbon paper and use the meterstick or measuring tape to accurately measure the horizontal distance the ball traveled before it hit the floor. Record the time interval, Δt, and horizontal distance, Δx, in the data table. Repeat this step two times.

12. Repeat steps 10 and 11. This time, roll the ball from the middle mark on the ramp three times. Each time, record the time and distance in your data table. Repeat again using the highest mark on the ramp.

13. After the last trial, choose NO to leave the pulse mode. Select RETURN TO MAIN, and then select QUIT to exit.

DATA TABLE

Distance between photogates (m), d						
Table height (m), Δy						

Launch point	Trial	Time (s), Δt	Horizontal velocity (m/s), v	Horizontal distance (m), Δx	Average velocity (m/s), v	Average distance (m), Δx
Low	1					
	2					
	3					
Medium	1					
	2					
	3					
High	1					
	2					
	3					

ANALYSIS

1. **Calculating** Use your time intervals and the distance between the photogates to calculate the horizontal velocity, v, of the ball for each trial. Velocity can be calculated as $v = d/\Delta t$. Enter these values in your data table.

2. **Averaging** Determine the average velocity and average horizontal distance for the ball traveling from each mark on the ramp, and enter them in your data table. Which mark produced the greatest average velocity?

3. **Graphing data** Use your calculator or a piece of graph paper to graph your data. Plot average velocity, v, as the independent variable and average horizontal distance, Δx, as the dependent variable. How does the graph compare to the sketch that you drew for item 2 of Developing the Model?

4. **Interpreting graphs** The graph of your data should suggest a linear relationship between the horizontal velocity, v, and horizontal distance, Δx, according to the following equation:

$$\Delta x = s_{obs} v$$

where s_{obs} is the slope of a line through the observational data. Use two points on the graph to calculate a slope, s_{obs}, using the following equation:

$$s_{obs} = \frac{\Delta x_2 - \Delta x_1}{v_2 - v_1}$$

5. **Applying theory** The equation in Chapter 3 of your textbook for the horizontal motion for projectile motion is $\Delta x = v_x \Delta t$. This equation is very similar to the equation for Δx in item 4 above, but now with a slope of Δt. Assume that this Δt is a theoretical slope, s_{th}, that corresponds to your observed slope, s_{obs}, and calculate s_{th} using the following equation:

$$s_{th} = \Delta t = \sqrt{\frac{2 \cdot \Delta y}{g}}$$

Use the table height for Δy and 9.81 m/s^2 for g.

6. **Evaluating results** To apply the laboratory model to the crash scene, you will have to evaluate the difference between the slope you derived from the model, s_{obs}, and the theoretical slope of the line, s_{th}. To quantify this difference, calculate a slope correction factor, k, using the following equation:

$$k = \frac{s_{obs}}{s_{th}}$$

7. **Interpreting results** What are the units for s_{obs} and s_{th}? What does s represent physically in both the model and the movie scene?

8. **Evaluating models** The difference between s_{obs} and s_{th} arises from factors in the real world that are not accounted for by the theory behind the equations of projectile motion. What are some of these factors that could account for the difference between these two values?

CONCLUSIONS

9. **Applying theory** Like the ball in your model, the car in the crash scene will act as a projectile once it is in the air. The same equations that describe the motion of the ball can be used to describe the predicted motion of the car. So, you can calculate a theoretical slope, s_{car}, of a line representing the relationship of velocity to horizontal distance for the car. Use the following equation:

$$s_{car} = \sqrt{\frac{2 \cdot \Delta y}{g}}$$

In this case, use the height of the cliff (112.0 m) for Δy.

10. **Making predictions** In evaluating your model, you found a difference between the theoretical slope and the observed slope. Assuming that many of the factors that affected the ball will also affect the car as it falls from the cliff, you can use the same correction factor, k, to make your prediction of the car's motion more realistic. Calculate the predicted slope for the car in the crash scene, s_{pred}, using the following equation:

$$s_{pred} = k(s_{car})$$

11. **Applying results** The director wants the crash to occur between 38 m and 42 m from the base of the cliff. Use the following equation to make a recommendation to the movie director for the initial velocity of the car, v_0.

$$v_0 = \frac{\Delta x}{s_{pred}}$$

Calculate both a maximum and a minimum horizontal speed for the car, and convert your results to units of miles per hour.

12. **Evaluating models** In what ways is the model of the ball rolling off a ramp similar to the actual scenario of a car driving off a cliff? In what ways is it different? How would these differences affect your recommendation to the director?

EXTENSIONS

1. **Making predictions** The film editor wants to limit the crash scene, including footage both of the car traveling toward the cliff and of the fall, to 8.00 seconds. How many seconds of the car traveling towards the edge of the cliff will the editor be able to include?

2. **Graphing** The director of photography would like to be able to predict the path that the car will take as it leaves the cliff. Use the equations from Chapter 3 of your textbook for horizontal and vertical displacement to determine the x- and y-coordinates for the car at 0.25 second intervals. Graph your results. Use your graph to estimate the time the car will take to hit the beach.

3. **Designing experiments** Later in the movie, there will be a scene involving a ski-jump competition. The ramp on the ski jump is tilted upward at the end so that the skiers leave the ramp at an angle. Design an experiment that could be used to model a ski jump on a movie set. If you have time and your teacher approves, carry out the experiment to determine how the angle may affect the value of s.

HOLT PHYSICS
Technology Lab A

Static and Kinetic Friction

Finding Safer Roofing Shoes

The manager of the RoofRite Company is getting ready to order new uniforms for the company's employees. RoofRite specializes in repairing and cleaning composition roofs on houses. In the past year, a couple of near accidents have occurred when employees have slipped while walking along slanted roofs.

The manager has also recently hired a number of high school students to work over the summer, and these students tend to weigh less than the company's older employees. The manager is concerned that the students' light weight may make them even more likely to slip.

To reduce the risk of accident, the manager has decided to include a new pair of shoes with the standard uniform. The company's uniform supplier offers a basic work shoe model, but these shoes are expensive, and the soles of the shoes do not grip roofing materials very well. The manager has asked you to explore the frictional characteristics of a variety of athletic and walking shoes.

Your goal is to find a pair that may offer more friction at less cost than the expensive work shoes that the uniform supplier has available. Your task is to use a force sensor to determine coefficients of static and kinetic friction for several pairs of shoes on composition roof shingles. You will also explore how weight may affect the force of friction on a shoe. You will use your data to select a pair of shoes to recommend to the manager.

OBJECTIVES

- **Develop** a method for determining coefficients of static and kinetic friction for shoes on roofing material.

- **Measure** forces of static and kinetic friction using a force sensor.

- **Determine** if forces of friction or coefficients of friction depend on weight.

- **Recommend** shoes that are suitable for people working on roofs.

MATERIALS

✔ graphing calculator with link cable

✔ CBL system

✔ PHYSICS application loaded in calculator

✔ Vernier force sensor

✔ two shoes (not from the same pair)

✔ composition roof shingles

✔ balance

✔ 500 g and 1 kg masses

✔ string

✔ tape

✔ graph paper

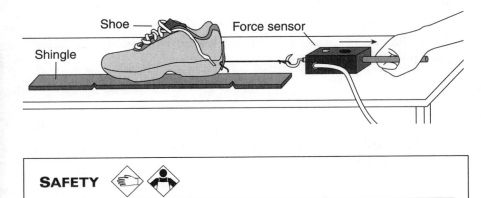

SAFETY

- Perform this experiment in a clear area. Attach masses securely. Falling, dropped, or swinging objects can cause serious injury.

- Tie back long hair, secure loose clothing, and remove loose jewelry to prevent them from getting caught.

DEVELOPING THE MODEL

The two basic kinds of frictional forces are static friction and kinetic friction. Each of these kinds of friction can be characterized using a coefficient of friction—μ_s for static friction and μ_k for kinetic friction. The best way to compare the frictional characteristics of different shoes is to determine these two coefficients of friction for each shoe.

Before starting this lab, do the following exercises:

1. Use a piece of tape to securely attach a piece of string to the heel of a shoe about one centimeter above the sole. Place the shoe on a table or on the floor in a clear area. Use the string to pull the shoe gently with a small horizontal force. Very gradually, taking at least one full second, increase the force until the shoe starts to slide. Once the shoe is moving, keep it moving at a constant speed for at least another second. Repeat this exercise several times, and feel how the force changes over time. Sketch a graph of the force you exerted over time.

2. Label the point on your graph where you felt you had to exert the strongest force. The force you exerted at that moment is the maximum force of static friction ($F_{s,max}$). What was happening to the shoe at that point?

3. Label the part of the graph in which the force you exerted was constant over time. This force is the force of kinetic friction (F_k). What was happening to the shoe during this time?

4. Describe how the graph you sketched might look different if someone's weight had been pressing down on the shoe.

5. What determines the magnitude of the normal force on a shoe resting on a flat surface? What determines the magnitude of the normal force on a shoe worn by a roofer walking across an inclined roof?

6. What force was responsible for causing the shoe to move in exercise 1? What force is responsible for causing a roofer's foot to slip along an inclined roof?

7. The coefficients of friction for static and kinetic friction can be calculated using the following equations:

$$\mu_s = \frac{F_{s,max}}{F_n} \text{ and } \mu_k = \frac{F_k}{F_n}$$

What do you think would happen to the value of μ_s and μ_k for the shoe/floor combination if the shoe had someone's foot pressing down on it? Explain your answer.

PROCEDURE

1. Prepare two sets of blank data tables like the ones shown in the Data Tables section. The first two tables will be for the static friction and kinetic friction, respectively, of the first shoe. The second two tables will be for the static friction and kinetic friction of the second shoe.

2. If the shoes do not already have strings attached, use tape to securely attach a piece of string to the heel of each shoe about one centimeter above the sole. Measure the mass of each shoe, and record your measurements in the appropriate data tables.

3. Connect the force sensor to the CH 1 input of the CBL unit. Use the black link cable to connect the CBL unit to the calculator. Firmly press in the cable ends.

4. Turn on the CBL unit and the calculator. Start the PHYSICS application and proceed to the MAIN MENU.

5. Set up the calculator and CBL for the force sensor.
   - Select SET UP PROBES from the MAIN MENU.
   - Select ONE as the number of probes.
   - Select FORCE from the SELECT PROBE menu.
   - Select STUDENT FORCE or DUAL-RANGE 5N as appropriate for your force sensor.
   - If using a dual-range force sensor, set the range switch to ±5N.
   - Confirm that the force sensor is attached to CH 1, and press [ENTER].
   - Select USE STORED from the CALIBRATION menu.

6. Zero the force sensor.
   - Hold the force sensor so that the sensitive axis is horizontal.
   - Select ZERO PROBES from the MAIN MENU.
   - Select CHANNEL 1 from the SELECT CHANNEL menu.
   - With no force applied to the force sensor, follow the instructions on the calculator screen to zero the sensor.

7. Set up the calculator and CBL for data collection.
   - Select COLLECT DATA from the MAIN MENU.
   - Select TIME GRAPH from the DATA COLLECTION menu.
   - Enter "0.05" as the time between samples, in seconds.
   - Enter "99" as the number of samples (the CBL will collect data for about 5 seconds).
   - Press [ENTER], then select USE TIME SETUP to continue. If you want to change the sample time or sample number, select MODIFY SETUP instead.

8. Place a piece of roof shingle on a table or on the floor in a clear area. Place one shoe on the shingle, and securely attach the free end of the string to the hook on the force sensor.

9. Hold the force sensor in position, ready to pull the handle, but with no tension in the string. Press [ENTER] to begin collecting data. Pull the handle of the force sensor gently away from the shoe with a small horizontal force. Very gradually, taking at least one full second, increase the force until the shoe starts to slide, then keep the shoe moving smoothly at a constant speed for at least another second. Another person may need to hold the shingle to keep it from sliding across the table.

10. Press [ENTER] to view your graph. Press [ENTER], and select YES to repeat the process as needed until you have a graph that reflects the desired motion, which includes pulling the shoe at a constant speed once it begins moving.

11. Trace along your final version of the graph using the cursor keys. The maximum value of the force of static friction occurs at the point when the shoe starts to slide. Read this value of the peak static friction force, and record the number in the appropriate data table. Draw a sketch of your final graph for later reference.

12. Next you will determine the average force of kinetic friction while the shoe was moving at constant velocity.
    - Press [ENTER] and select NO to return to the MAIN MENU.
    - Select ANALYZE from the MAIN MENU.
    - Select STATS/INTEGRAL from the ANALYZE MENU.
    - Select STATISTICS from the STATS/INTEGRAL menu.
    - Select CHANNEL 1 from the SELECT GRAPH menu.
    - Select a portion of the force graph for averaging. Using the cursor keys, move the lower bound cursor to the left side of the approximately constant-force region. Press [ENTER].
    - Now select the other edge. Move the cursor to the right edge of the approximately constant-force region. Press [ENTER].
    - When prompted, press [ENTER] to continue. Read the mean force from the calculator. Record this kinetic friction value in the appropriate data table.
    - Press [ENTER] to return to the MAIN MENU.

13. Repeat steps 7–12 two more times with the same shoe. Record the results as trials 2 and 3 in your data tables. Average the results of the peak static friction and the kinetic friction for all three trials, and record the average values in your data tables.

14. Now insert a 500 g mass (or several masses totaling 500 g) into the shoe. Repeat steps 7–13, and record all values in your data tables. Repeat steps 7–13 once more with a 1 kg mass inside the shoe.

15. Repeat steps 7–14 with a second shoe, and record all values in the second pair of data tables.

DATA TABLES

Shoe: _____

Mass of shoe: _____ **kg**

Total mass (kg)	Normal force (N)	Peak static friction (N)			Average peak static friction (N)
		Trial 1	Trial 2	Trial 3	
				μ_s (slope)	

Shoe: _____

Mass of shoe: _____kg

Total mass (kg)	Normal force (N)	Kinetic friction (N)			Average peak static friction (N)
		Trial 1	Trial 2	Trial 3	
				μ_k (slope)	

ANALYSIS

1. **Interpreting graphs** On each sketch of the graphs you generated in the Procedure, label the portion of the graph corresponding to the time the shoe was at rest, the point when the shoe just started to move, and the portion corresponding to the time when the shoe was moving at constant speed.

2. **Interpreting graphs** Based on the graphs you generated, which is greater: the maximum force of static friction or the force of kinetic friction?

3. **Interpreting graphs** Based on the graphs you generated, would you expect the coefficient of static friction to be greater than, less than, or the same as the coefficient of kinetic friction? Explain.

4. **Calculating** Calculate the magnitude of the normal force, F_n, of the table on each shoe, alone and with each combination of added masses. To do so, use the equation $F_n = F_g = mg$. Record the values in the appropriate data tables.

5. **Graphing data** Plot graphs of the average peak force of static friction (y-axis) versus the normal force (x-axis) for each shoe.

6. **Interpreting graphs** According to the equation $F_{s,max} = \mu_s F_n$, the slope of the line in each graph in item 5 is the coefficient of static friction, μ_s, between the shoe and the shingle. Use the graphs of static friction versus normal force to determine μ_s for each shoe, and record the results in your data tables.

7. **Graphing data** Plot graphs of the average force of kinetic friction (y-axis) versus the normal force (x-axis) for each shoe.

8. **Interpreting graphs** According to the equation $F_k = \mu_k F_n$, the slope of the line in each graph in item 7 is the coefficient of kinetic friction, μ_k, between the shoe and the shingle. Use the graphs of kinetic friction versus normal force to determine μ_k for each shoe, and record the results in your data tables.

CONCLUSIONS

9. **Analyzing results** According to your data, do either the maximum force of static friction or the force of kinetic friction depend on the weight acting on a shoe? Explain how this does or does not reflect the equations for the forces of static and kinetic friction.

10. **Analyzing results** According to your data, do the coefficients of static and kinetic friction depend on the weight acting on a shoe? Explain.

11. **Analyzing results** Should the manager of RoofRite be more concerned about the risk of slipping for smaller, lighter workers than for heavier workers? Explain why or why not.

12. **Reaching conclusions** Of the two kinds of shoes you tested, which one would you recommend to the manager of the roofing company? Explain why you would recommend that kind of shoe over the other kind.

13. **Comparing results** Compare the shoes you tested with the shoes that others in your class tested. In general, what are some of the common characteristics of shoes with high coefficients of friction? Of all the shoes tested in your class, which would you recommend to the manager of the roofing company?

EXTENSIONS

1. **Evaluating safety** Describe the potential safety risk for a roofer wearing shoes with a high μ_s but a low μ_k. What advantages are there for a roofer to wear shoes in which the μ_s and the μ_k are nearly the same?

2. **Making predictions** Predict how increasing the surface area of a shoe's sole would affect forces of friction and coefficients of friction. Devise an experiment that would test your hypothesis. If you have time, and if your teacher approves your plan, perform this experiment.

3. **Extending results** Use the force sensor to measure the maximum force of static friction for a shoe on an inclined roof shingle. Find the angle that causes a shoe to start to slide. Calculate the coefficient of friction, and compare it to the value you obtain when the shingle is lying flat.

4. **Extending research** Some sport sandals, climbing shoes, and hiking boots have "sticky" rubber soles that are reported to have very high coefficients of friction. Research these types of materials and prepare a report that explains how they are made and that discusses new research on similar materials.

Technology Lab B

Air Resistance

Planning a Wilderness Supply Drop

You are part of a technical support crew that will be delivering delicate weather equipment to scientists in remote wilderness areas of Alaska. The equipment will be delivered by plane, using supply boxes attached to parachutes. The boxes are designed to withstand impact at velocities up to 1.5 m/s. The heaviest piece of equipment you must deliver weighs about 55 kg.

You have been asked to recommend an appropriate parachute (one that offers adequate air resistance) and to determine a maximum possible load per drop. To do so, you must first determine the mathematical model that best describes the relationship between the mass of an object, the air resistance on the object, and the terminal speed the object reaches.

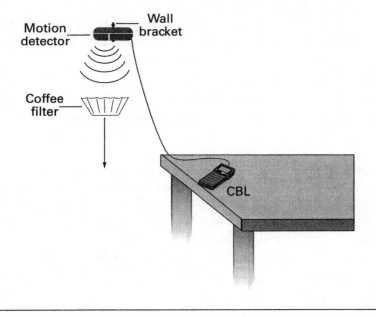

OBJECTIVES

- **Develop** a physical model that simulates an object falling with a parachute.

- **Analyze** the relationship between mass and terminal speed of a falling object using data from the model.

- **Evaluate** two mathematical models relating air resistance to terminal speed using data generated in the simulation.

- **Calculate** an air resistance factor for the model.

- **Predict** the required air resistance factor for the parachute and the maximum allowable load for the drop.

MATERIALS

- ✔ graphing calculator with link cable
- ✔ CBL system
- ✔ PHYSICS application loaded in calculator
- ✔ Vernier motion detector
- ✔ 5 basket-style coffee filters
- ✔ balance
- ✔ graph paper

SAFETY

- Perform this experiment in a clear area. Attach masses securely. Falling, dropped, or swinging masses can cause serious injury. Use caution when standing on ladders or chairs.

- Tie back long hair, secure loose clothing, and remove loose jewelry to keep them from getting caught in moving or rotating parts.

DEVELOPING THE MODEL

To determine a mathematical model for air resistance in the laboratory, you can drop an object that is similar to a parachute and use a motion detector to collect data as it falls. To simulate a parachute in this experiment, you will use coffee filters dropped right-side-up. The factors affecting the upward force of air resistance on the filters will be combined into a single number called the "air resistance factor."

Before starting this activity, answer the following questions:

1. For a falling object, what is the mathematical expression that describes the downward force acting on the object?

2. When an object reaches terminal speed, what is the net force acting on it? Explain your answer.

3. Sketch a graph of velocity versus time for a parachute carrying a small load as it falls through the air. Label the point on the graph at which terminal speed occurs. On the same set of axes, sketch a graph of a parachute with a much larger weight attached. How does the weight of the load affect the terminal speed the parachute reaches?

4. One possible mathematical model for the air resistance on a parachute is that the air resistance is directly proportional to the speed ($F_R = -kv$). Assuming this is the case, find an expression for terminal speed, v_T, in terms of g, m, and k, where g is the free-fall acceleration, m is mass, and k is a constant air resistance factor. (Hint: Set $F_R = -F_g$ and solve for v.)

5. Another possible mathematical model for the air resistance on a parachute is that the air resistance is directly proportional to the square of the speed ($F_R = -kv^2$). Assuming this is the case, find an expression for terminal speed, v_T, in terms of g, m, and k.

PROCEDURE

1. Mount the motion detector at least two meters above the ground using a bracket on the wall or ceiling. Orient the motion detector so that it faces straight down toward the floor.

2. Connect the motion detector to the SONIC port of the CBL unit. Use the black link cable to connect the CBL unit to the calculator. Firmly press in the cable ends.

3. Set up the calculator and CBL for the motion detector. Start the PHYSICS application and proceed to the MAIN MENU.

   • Select SET UP PROBES from the MAIN MENU.

   • Select ONE as the number of probes.

   • Select MOTION from the SELECT PROBE menu.

4. Set up the calculator and CBL for data collection.

   • Select COLLECT DATA from the MAIN MENU.

   • Select TIME GRAPH from the DATA COLLECTION menu.

   • Enter "0.03" as the time between samples, in seconds.

   • Enter "99" as the number of samples (the CBL will collect data for about three seconds).

- Press ENTER, then select USE TIME SETUP to continue. If you want to change the sample time or sample number, select MODIFY SETUP instead.

5. Determine the mass of a single coffer filter and record the mass in a data table like the one shown in the Data Table section.

6. Make sure the area around you is free of obstructions. Hold the coffee filter about 0.5 m under the motion detector. Do not hold the filter closer than 0.4 m. Press ENTER to begin data collection. When the motion detector begins to click, release the coffee filter directly below the motion detector so that it falls toward the floor. Move your hands out of the beam of the motion detector as quickly as possible so that only the motion of the filter is recorded.

7. View a graph of distance versus time.

- Press ENTER to view the SELECT GRAPH menu.

- Select DISTANCE to view a distance-time graph.

- If the motion of the filter was too erratic to get a smooth graph, repeat the measurement. With practice, you should be able to release the filter so that it falls almost straight down with little sideways motion.

- Press ENTER and select NEXT.

- Select NO to continue, or YES to repeat data collection.

8. The speed of the coffee filter can be determined from the slope of the distance-time graph. At the start of the graph, there should be a region of increasing slope (increasing speed), and then the plot should become linear because the filter was falling with a constant or terminal speed (v_T) during that time. To fit a line to only the linear region:

- Select ANALYZE from the MAIN MENU.

- Select SELECT REGION from the ANALYZE MENU.

- Select DISTANCE from the SELECT GRAPH menu.

- Move the flashing cursor with the cursor keys to the left edge of the linear region corresponding to the filter in motion at constant speed, and press ENTER.

- Move the flashing cursor to the right edge of the linear region, and press ENTER.

- View your abbreviated graph by selecting DISTANCE from the SELECT GRAPH menu. You should see only the linear region.

- Press ENTER, and select NEXT.

- Select ANALYZE from the MAIN MENU.

- Select CURVE FIT from the ANALYZE.

- Select LINEAR L_1, L_4 from the CURVE FIT menu.

- Record the slope under *Terminal speed, v_T (m/s)* in your data table.

- Press ENTER to see the fit along with your data.

- Press ENTER to return to the MAIN MENU.

9. Repeat steps 5–8 for nested stacks of two, three, four, and five coffee filters. (Optionally extend to six, seven, and eight filters, but be sure to allow sufficient falling distance so that a terminal speed will be reached.) Record all data in your data table.

DATA TABLE

Number of filters	Total mass, m (kg)	Terminal speed, v_T (m/s)	v_T^2 (m²/s²)	Air resistance factor, k (kg/m)
1				
2				
3				
4				
5				
			Average =	

ANALYSIS

1. **Graphing Data** Use your calculator or graph paper to plot terminal speed, v_T, versus mass, m, for the five trials. Be sure to scale the axes from the origin (0,0). Draw a line through your data that also goes through the origin. Does your data fit a linear model? Explain why or why not.

2. **Calculate** Square each terminal speed in the data table, and record the results under the heading v_T^2 (m²/s²) in your data table.

3. **Graphing Data** On a separate graph, plot terminal speed squared, v_T^2, versus mass, m. Again, scale the axes through the origin. Does this seem to be a better fit than the linear model? Explain why or why not.

4. **Interpreting graphs** Based on your data and graphs, which mathematical model best represents the relationship between the force of air resistance and the speed of the coffee filters? (Choose a or b.)

   a. $F_R = -kv$ (linear model)

   b. $F_R = -kv^2$ (quadratic model)

5. **Evaluating results** Calculate an air resistance factor, k, for each of the coffee filter trials. If you found that your data fit a linear model better, use the following equation:

$$k = \frac{mg}{v_T}$$

   If you found that your data fit a quadratic model best, use the following equation instead:

$$k = \frac{mg}{v_T^2}$$

   Record the values for k in your data table, then calculate an average air resistance factor for all the trials combined.

CONCLUSIONS

6. **Applying the model** Assume that air resistance on the parachute follows the same mathematical model as the coffee filters. If the support team decided to deliver equipment in batches weighing 200 kg, what would the minimum required air resistance factor for the parachute be? Remember that the impact speed can be no greater than 1.5 m/s. (Hint: Use one of the equations from step 4 in the Analysis.)

7. **Making predictions** Your current parachutes have air resistance factors of 160 kg/m. What is the maximum load you can deliver without exceeding the impact speed of 1.5 m/s? The heaviest single piece of equipment that must be delivered weighs 55 kg. Will your team be able to use the current parachute to safely deliver this equipment? (Hint: Use one of the equations from step 5 in the Analysis and solve for mass, m.)

8. **Applying the model** On a recent drop, one of the parachutes did not open. The supply box, with a mass of 200 kg and an air resistance factor of 0.8, continued to accelerate without its parachute. What was the terminal speed that was reached by the supply box?

9. **Evaluating models** If a parachute and a coffee filter each had the same cross-sectional area, which would offer more air resistance? Explain why you think so. Do you think the differences between them would result in an entirely different mathematical model?

EXTENSIONS

1. **Evaluating models** Design a small parachute and have your teacher approve your design. Use the motion detector to analyze the air resistance and terminal speed as the weight suspended from the parachute increases. Determine whether or not the parachute uses the same mathematical model as the coffee filter.

2. **Developing models** The air resistance factor, k, used in this experiment combines several factors into a single constant. When physicists study air resistance, they sometimes use the following equation to summarize the force of air resistance:

$$F_R = (1/2)^* C_D^* A^* r^* v^2$$

where C_D is a *drag coefficient* based on the overall shape of an object (a perfect sphere has a $C_D = 0.5$), A is the cross-sectional area in m^2 of the falling object, and r refers to the characteristics of the medium through which the object is falling (in air, r is equal to 1.2 kg/m^3). Use this equation and your data to calculate the drag coefficient, C_D, for the coffee filter.

3. **Applying the model** Look up the drag coefficients for several standard parachute designs, and calculate the size parachute required with each design to safely deliver a 55 kg load with a terminal velocity no greater than 1.5 m/s. Use the equation in item 2 above.

4. **Applying the model** A 65 kg stuntwoman jumps from a plane using a giant coffee filter as a parachute. Use the drag coefficient, C_D, that you calculated in item 2 above to determine the necessary cross-sectional area of the giant filter if she does not want to exceed a speed of 3 m/s while falling.

Technology Lab

Loss of Mechanical Energy

OBJECTIVES

- **Measure** the change in the kinetic and potential energy as a ball moves up and down in free fall.
- **Graph** potential energy, kinetic energy, and total energy.
- **Analyze** the graph to determine how much kinetic energy is lost.
- **Reach conclusions** regarding the amount of energy possessed by the volleyball as it fell in the neighbor's driveway.

MATERIALS

- ✔ graphing calculator with link cable
- ✔ CBL system
- ✔ PHYSICS application loaded in calculator
- ✔ Vernier motion detector
- ✔ wire basket
- ✔ volleyball or other ball

The Case of the '65 Mustang

One day, as you are playing volleyball with a group of friends in your driveway, a stray serve accidentally goes over the fence to the neighbor's driveway. When you go to retrieve the volleyball, your neighbor points out a dent in the roof of his classic 1965 Mustang, claiming that the volleyball hit the car. You apologize to your neighbor for the ball going over the fence, but explain that you know the volleyball didn't hit the car because you heard it land on the pavement.

The next day, the neighbor presents you with an $1800 repair bill. Not wanting the dispute to get out of hand, you suggest turning to the local neighborhood organization for arbitration. Your neighbor agrees, and within a week the president of the organization contacts you. Your neighbor has already presented his case to the president, using the following:

- A passage from a physics textbook stating that mechanical energy is conserved during projectile motion, even though it may be converted from kinetic energy to potential energy and back to kinetic energy again.
- Calculations estimating that the ball left the server's hand (about 1.5 m in height) with an initial vertical velocity component of 8.29 m/s, rose to a height of 5 m (passing just under the overhanging branches of a tree), and hit the roof of the car (also at 1.5 m) with a downward velocity of 8.29 m/s.
- Calculations showing that, based on a ball mass of 0.5 kg and an impact velocity of 8.29 m/s, the kinetic energy of the ball at the time of impact was 19.58 J.
- Testimony by a materials engineer that any impact by an object with kinetic energy greater than 18 J could dent the roof of the Mustang.

Unfortunately, the evidence appears to be in strong support of your neighbor's claim. Your only chance to defend yourself is to present an argument that the ball possessed less kinetic energy when it fell than when it was served so that, even according to the neighbor's own data, the ball could not have damaged the car.

SAFETY

- Perform this experiment in a clear area. When tossing balls for the experiment, throw the balls straight up and catch them as they approach the ground. Do not throw balls around the room.

- Tie back long hair, secure loose clothing, and remove loose jewelry to keep them from getting in the way.

DEVELOPING THE MODEL

You have decided that the best way to counter your neighbor's argument is to collect and present some real data. To simulate the volleyball going over the fence, you will toss a ball straight up over a motion detector, as shown in **Figure 5-1.** The motion detector will measure the distance between the detector and the ball at regular time intervals. The distance data can be used to calculate velocity, kinetic energy, and total mechanical energy at different points in the trajectory of the ball. You will then be able to determine if the ascending velocities and descending velocities, and the corresponding kinetic energies, are the same. If they are not the same, you will be able to determine by how much they differ.

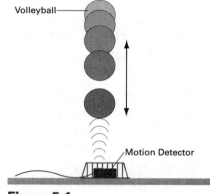

Figure 5-1

Before you begin this activity, answer the following questions:

1. Describe what happens to the potential energy of a ball that is thrown upward and then falls back to the ground. What values must be known to calculate potential energy?

2. Describe what happens to the kinetic energy of a ball that is thrown upward and then falls back to the ground. What values must be known to calculate kinetic energy?

3. Sketch a graph of velocity versus time for a ball that is thrown straight upward and then falls back to the ground. Use upward as the positive direction. Would you expect the velocity—time graph to show a straight line? Why or why not?

4. Sketch a graph of vertical distance from the ground (height) versus time for a ball that is thrown straight upward and then falls back to the ground. Would you expect the graph to be symmetrical? Why or why not?

5. What factors may contribute to a loss of energy as a ball moves through the air? Do you think those factors would have a measurable effect in the model?

PROCEDURE

1. Measure the mass of the ball and record the mass in a data table like the one shown in the Data Table section.

2. Connect the Vernier motion detector to the SONIC port of the CBL unit. Place the detector on the floor and protect it by placing a wire basket over it. Place the CBL unit a safe distance away from the basket, so it will not be hit by the ball as it falls. Use the black link cable to connect the CBL unit to the calculator. Firmly press in the cable ends.

3. Set up the calculator and CBL for the motion detector. Start the PHYSICS application and proceed to the MAIN MENU.

- Select SET UP PROBES from the MAIN MENU.
- Select ONE as the number of probes.
- Select MOTION from the SELECT PROBE menu.

4. Set up the calculator and CBL for data collection.

- Select COLLECT DATA from the MAIN MENU.
- Select TIME GRAPH from the DATA COLLECTION menu.
- Enter "0.05" as the time between samples, in seconds.
- Enter "80" as the number of samples (the CBL will collect data for four seconds).
- Press ENTER, then select USE TIME SETUP to continue. If you want to change the sample time or sample number, select MODIFY SETUP instead.

5. In this step, you will toss the ball straight upward above the motion detector and let it fall back toward the motion detector. Make sure the area around the motion detector is clear and that all cables are out of the way. Hold the ball in both hands directly above and about 0.5 m from the motion detector. Have one of your lab partners press ENTER to begin data collection. After you hear the motion detector begin to click, toss the ball straight up into the air. Pull your hands away so the motion detector does not detect them instead of the ball. Either you or another one of your lab partners should catch the ball when it returns to a point about 0.5 m above the motion detector. Hold the ball steadily in place until the motion detector stops collecting data (until it stops clicking).

6. Press ENTER to proceed to the SELECT GRAPH menu. Select the DISTANCE graph. Examine the graph. Repeat step 5 if your distance versus time graph does not show a smooth region of changing distance. Check with your teacher if you are not sure whether you need to repeat the data collection. To repeat, press ENTER, select NEXT, and then select YES from the REPEAT? menu.

DATA TABLE

Mass of ball: _____ kg

Position	Time (s)	Height (m)	Velocity (m/s)	PE (J)	KE (J)	ME (J)
Just after release						
Between release and top						
Top of path						
Between top and catch						
Just before catch						

ANALYSIS

1. **Isolating data** In order to analyze only the desired portion of the data, you must discard all other data:

   - Press [ENTER] to return to the SELECT GRAPH menu.

   - Select NEXT from the SELECT GRAPH menu.

   - Select NO from the REPEAT? menu.

   - Select ANALYZE from the MAIN MENU.

   - Select SELECT REGION from the ANALYZE MENU.

   - Select VELOCITY from the SELECT GRAPH menu.

   - Using the cursor keys, move the flashing cursor to the beginning of the straight-line, negative-slope region. This portion corresponds to free fall.

   - Press [ENTER] to select the lower bound.

   - Using the cursor keys, move the flashing cursor to the end of the straight-line section corresponding to free fall.

   - Press [ENTER] to select the upper bound.

2. **Graph tracing** To explore the kinetic and potential energy of the ball at various points during the motion, you can trace along the distance-time and velocity-time graphs using the cursor keys. First trace along the velocity-time graph, and record the velocity at five different times in your data table. Use the point just after the ball was released into free fall, the point where the ball was at the top of the path, the point just before the ball was caught, and two points approximately halfway between the other points. Sketch this graph onto a piece of paper for later reference.

3. **Graph tracing** Press [ENTER], and choose DISTANCE to see the distance-time graph. Trace along this graph to the same five times you used in the previous step, and record the corresponding distances in your data table. Because the motion detector was on the ground and the ball moved straight up and down over it, these distances are in effect the height of the ball at those times. Sketch this graph for later reference.

4. **Applying theory** The remaining calculations will require manipulating the data outside of the PHYSICS application. Press [ENTER], NEXT, and then QUIT to exit the application. Then, for each of the five points in your data table, calculate the potential energy ($PE = mgh$), kinetic energy ($KE = \frac{1}{2}mv^2$), and total mechanical energy ($ME = PE + KE$).

5. **Generating lists** The PHYSICS application stores distance data in the calculator's list table L_4, and velocity data in the list table L_5. Now you can use the data in these lists to calculate the energy of the ball at every data point, and store the results in additional list tables on the calculator.

   - To calculate the ball's kinetic energy at every point and store the values in the list L_2, type 0.5 [×] (*mass of ball*) [×] [2nd] L5 [^] 2 [STO▸] [2nd] L2 [ENTER]. For *mass of ball*, enter the mass of the ball that you measured in step 1 of the Procedure.

   - To calculate the ball's gravitational potential energy at every point and store the values in the list L_3, type 9.8 [×] (*mass of ball*) [×] [2nd] L4 [STO▸] [2nd] L3 [ENTER].

- To calculate the ball's total mechanical energy at every point and store the values in the list L_6, type [2nd] L2 [+] [2nd] L3 [STO▶] [2nd] L6 [ENTER].

6. Generating graphs Plot a combined graph of the kinetic energy, potential energy, and total mechanical energy of the ball using the following steps.

- Press [2nd] STAT PLOT and select Plot 1.

- Use the arrow keys to position the cursor on each of the following Plot1 settings. Press [ENTER] to select any of the settings you change: Plot1 = On, Type = ⌊⋯⌋, Xlist = L_1, Ylist = L_2, and Mark = □. This is the kinetic energy.

- Press [2nd] STAT PLOT and select Plot 2.

- Use the arrow keys to position the cursor on each of the following Plot2 settings. Press [ENTER] to select any of the settings you change: Plot2 = On, Type = ⌊⋯⌋, Xlist = L_1, Ylist = L_3, and Mark = +. This is the potential energy.

- Press [2nd] STAT PLOT and select Plot 3.

- Use the arrow keys to position the cursor on each of the following Plot3 settings. Press [ENTER] to select any of the settings you change: Plot3 = On, Type = ⌊⌒⌋, Xlist = L_1, Ylist = L_6, and Mark = •. This is the total energy.

- Press [ZOOM] and then select ZoomStat (use cursor keys to scroll to ZoomStat) to draw a graph with the x- and y-ranges set to fill the screen with data.

- Sketch this graph onto a piece of paper for later reference. Be sure to mark which curve corresponds to which type of energy (kinetic energy is marked with □ symbols, potential energy with + symbols, and total mechanical energy with a solid line).

CONCLUSIONS

7. Comparing How do the graphs of distance versus time and velocity versus time that were generated during the experiment (steps 2 and 3 of the Procedure) compare to the graphs you sketched before doing the experiment (items 3 and 4 of Developing the Model)? If the graphs differ significantly, explain why they are different.

8. Interpreting graphs According to your energy graph, did the total mechanical energy of the ball change significantly over time? If so, how did it change? What features of the graph support your conclusion?

9. Calculating percentages Calculate the difference between the total mechanical energy just after the ball was released, ME_i, and the total mechanical energy just before the ball was caught, ME_f. This difference represents the amount of mechanical energy lost during the flight of the ball. Use the following equation to calculate the total energy loss as a percentage:

$$percent\ energy\ loss = \frac{ME_i - ME_f}{ME_i} \times 100\%$$

10. **Applying results** According to your neighbor's estimates, the volleyball that went over the fence had an initial kinetic energy of 19.6 J when it first left the hand of the server. Assuming a 15 percent loss of mechanical energy during the flight of the ball, estimate the amount of kinetic energy the volleyball would have had when it reached the roof of the car (at approximately the same height as the server's hand).

11. **Reaching conclusions** The materials engineer stated that a ball would need to have at least 18 J of kinetic energy in order to dent the roof of the Mustang. Based on your results, could the volleyball have dented the roof?

EXTENSIONS

1. **Making predictions** When the volleyball flew over the wall, it had a horizontal component to the velocity in addition to the vertical component. The model only accounted for vertical motion. If this point were brought out to the arbitration committee, how might it affect their conclusion?

2. **Making predictions** How would your data have been different if you had used a very light ball, such as a beach ball. If you have time, test your hypothesis.

3. **Thinking critically** What would happen to your results if you entered the wrong mass for the ball in this experiment? Explain why or why not your conclusions may be different.

4. **Designing experiments** If you have time to perform additional experiments, use the motion sensor to determine the loss of kinetic energy as a ball bounces. Mount the motion detector on the ceiling or on a high wall bracket, and point the detector downward so it can follow a ball through several bounces. Test several different types of balls and compare their bounce characteristics. Perform these experiments in a clear area with a level floor.

Technology Lab

Impulse and Momentum

Demonstrating a High-Impulse Jet Landing

Military pilots sometimes have to land jets on short runways located on aircraft carriers at sea. Because of the limited length of the aircraft carrier deck, spring lines are used to slow the jets down so that the jets are able to come to a complete stop before reaching the end of the deck. A spring line is a heavy elastic cord that is stretched across the runway in front of the landing jet. After traveling a short distance down the runway, the jet hits the spring line and additional force is exerted on the jet. The increased force decreases the amount of time required for the jet to reduce its momentum to zero, thus shortening the stopping distance.

The dependence on a spring line is a source of anxiety for trainees landing on an aircraft carrier for the first time. To alleviate some of the anxiety, new recruits attend a seminar that explains how spring lines work. You have been hired as the seminar instructor. Your task is to provide a model of how the spring line works and to explain the relationship between force, the change in momentum, and the length of time force is applied.

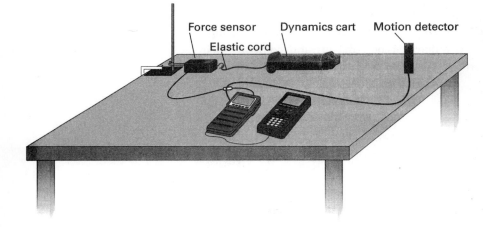

Force sensor   Dynamics cart   Motion detector
Elastic cord

OBJECTIVES

- **Develop** a model that demonstrates the impulse-momentum theorem.
- **Measure** momentum change and impulse using the model under various conditions.
- **Relate** the model and the impulse-momentum theorem to military pilots and the stopping distances of jets.

MATERIALS

- ✔ graphing calculator with link cable
- ✔ CBL system
- ✔ PHYSICS application loaded in calculator
- ✔ Vernier motion detector
- ✔ Vernier force sensor and CBL adapter
- ✔ support stand and clamp
- ✔ dynamics cart
- ✔ elastic cord
- ✔ string
- ✔ 500 g mass

SAFETY

- Perform this experiment in a clear area. Attach masses securely. Falling, dropped, or swinging masses can cause serious injury.

- Tie back long hair, secure loose clothing, and remove loose jewelry to keep them from getting caught in moving or rotating parts.

DEVELOPING THE MODEL

You will use a low-friction dynamics cart rolling on a tabletop to simulate a jet on the deck of an aircraft carrier. An elastic cord attached to the cart will act like the spring line that is stretched across the runway. Using a force sensor and a motion detector, you will be able to simultaneously measure both the changing velocity of the cart and the forces acting on the cart. You can use these measurements to calculate impulse and momentum changes on the cart.

Before starting this lab, answer the following questions:

1. The impulse-momentum theorem relates impulse (the average force applied to an object times the length of time the force is applied) and the change in momentum of the object using the following equation:

$$\mathbf{F}\Delta t = \Delta \mathbf{p} = m\mathbf{v_f} - m\mathbf{v_i}$$

For the jet landing on a runway, the final speed, v_f, is 0 m/s. How will this affect the above equation?

2. Sketch a graph of the force acting on a jet from the moment the jet hits the spring line to the moment in which the spring line is extended to its maximum and the jet has a speed of 0 m/s. Label the point at which the spring line begins to stretch and the point at which the spring line is stretched the most.

3. For a jet landing on the deck of an aircraft carrier, what forces are acting to slow the jet down? Compare those forces with the forces acting on a cart attached to an elastic cord. How is the model different from an actual jet landing?

4. For engineers trying to design a spring line for the deck of an aircraft carrier, what factors would they have to consider? How could those factors be tested in the model?

PROCEDURE

1. Place your dynamics cart on a lab balance with the wheels up so it does not roll off. Measure the mass of the cart and record the value in a data table like the first one shown in the Data Tables section.

2. Place the cart on a level tabletop. Attach the elastic cord to the cart and then the cord to a piece of string about 0.5 m in length. Tie the string to the force sensor, and clamp the force sensor to the support stand on one end of the table. Secure the support stand using a table clamp. The string and cord, when taut, should be horizontal and in line with the cart's motion.

3. Place the motion detector in line with the cart and the force sensor on the opposite side of the cart, as shown in the figure on the previous page. The detector should directly face the cart and should have a clear view of the cart throughout the cart's motion. When the cord is stretched to maximum extension, the cart should not be closer than 0.4 m to the detector.

4. Using the adapter cable, connect the force sensor to CH 1 of the CBL. Connect the motion detector to the SONIC input of the CBL. Use the black link cable to connect the CBL unit to the calculator. Firmly press in the cable ends.

5. Practice pushing the cart so it rolls toward the motion detector, stretches the elastic band, and gently bounces back toward your hand. Unlike an actual

aircraft landing, this should be a gentle motion. The force sensor must not shift, and the cart must move in a straight line. Arrange the cord and string so that when they are slack they do not interfere with the cart's motion. You may need to guide the string by hand, but do not apply any unwanted forces to the cart or force sensor. Keep your hands clear of the space between the cart and the motion detector. Be prepared to stop the cart before it hits the motion detector if the cord stretches too far or breaks.

6. Turn on the CBL unit and the calculator. Start the PHYSICS application and proceed to the MAIN MENU.

7. Set up the calculator and CBL for the motion detector and the force sensor.
   - Select SET UP PROBES from the MAIN MENU.
   - Select TWO as the number of probes.
   - Select MOTION from the SELECT PROBE menu.
   - Select FORCE from the SELECT PROBE menu.
   - Select STUDENT FORCE, DUAL-RANGE 5N, or DUAL-RANGE 10N as appropriate for your force sensor.
   - If using a dual-range force sensor, set the range switch to ±5N or ±10N as appropriate for your sensor.
   - Confirm that the force sensor is connected to CHANNEL 1, and press ENTER.
   - Select USE STORED from the CALIBRATION menu.

8. Zero the force sensor.
   - Remove all force from the force sensor.
   - Select ZERO PROBES from the MAIN MENU.
   - Select CHANNEL 1 from the SELECT CHANNEL menu.
   - Follow the instructions on the calculator screen to zero the sensor.

9. Set up the calculator and CBL for data collection.
   - Select COLLECT DATA from the MAIN MENU.
   - Select TIME GRAPH from the DATA COLLECTION menu.
   - Enter "0.02" as the time between samples in seconds.
   - Enter "150" as the number of samples (data will be collected for three seconds). Note: if you are using an older calculator, such as a TI-82 or TI-83, set "99" as the number of samples instead.
   - Press ENTER, and select USE TIME SETUP to continue. If you want to change the sample time or sample number, select MODIFY SETUP instead.

10. Press ENTER to begin collecting data. As soon as you hear the motion detector start clicking, roll the cart.
    - When you have finished collecting data, press ENTER and select DISTANCE from the SELECT GRAPH menu.
    - Confirm that the motion detector detected the cart throughout its travel. If there is a noisy or flat spot near the time of closest approach, then the motion detector was too close to the cart. If that is the case, move the motion detector away from the cart, and repeat your data collection.

- Press [ENTER] and select CHANNEL 1 from the SELECT GRAPH menu.
- Inspect the graph of force versus time. If the peak is flattened, then the applied force is too large. If that is the case, repeat your data collection with a lower initial speed.
- If you are satisfied with the graphs, make a sketch of the graph of force versus time for later reference. Use the cursor keys to note the time at which the force was initially exerted and the time at which the force was the greatest.
- Press [ENTER], then select NEXT.
- Select YES to collect data again, or select NO to go on.

11. Once you have completed a trial with good distance and force graphs, proceed to steps 1–3 of the Analysis. When you are finished with those steps, return to this point in the Procedure to collect additional data.

12. Perform a second trial by repeating steps 9–11, including the required steps of the Analysis. Record all data and results in your data table.

13. Repeat steps 9–12 again, this time using an elastic cord of a different material or attaching two elastic bands side by side. Record all data in your data table. When you are finished collecting all data, move on to steps 4–6 of the Analysis and the Conclusions.

DATA TABLES

Mass of cart: _____ kg

Cord	Trial	Stopping distance, d (m)	Final velocity, v_f (m/s)	Initial velocity, v_i (m/s)	Change in velocity, Δv (m/s)	Average force, F (N)	Duration of impulse, Δt (s)	Impulse (N•s)
elastic 1	1		0					
	2		0					
elastic 2	1		0					
	2		0					

Cord	Trial	Impulse, $F\Delta t$ (N•s)	Change in momentum, Δp (kg•m/s or N•s)	Percent difference between $F\Delta t$ and Δp
elastic 1	1			
	2			
elastic 2	1			
	2			

ANALYSIS

1. **Graph tracing** Use the following steps to determine the initial velocity before the impulse.
   - Select ANALYZE from the MAIN MENU.
   - Select STATS/INTEGRAL from the ANALYZE MENU.

- Select STATISTICS from the STATS/INTEGRAL MENU.

- Select VELOCITY from SELECT GRAPHS menu. Sketch the graph of velocity versus time for later reference.

- Now select the appropriate portion of the velocity graph for averaging. Using the cursor keys, move the lower-bound cursor to the left edge of the region where the velocity is negative and nearly constant. Press [ENTER].

- Now move the upper-bound cursor to the right edge of the region where the velocity is negative and nearly constant. Press [ENTER].

- When prompted, press [ENTER], and read the average velocity before the cart starts to slow down. Record the value as *initial velocity*, v_i, in your data table. The velocity values are negative because the cart is moving toward the motion detector during that time.

- Press [ENTER] to return to the MAIN MENU.

2. **Graph tracing** Use the following steps to determine the average applied force and the time duration in which the force was applied.

   - Select ANALYZE from the MAIN MENU.

   - Select STATS/INTEGRAL from the ANALYZE MENU.

   - Select STATISTICS from the STATS/INTEGRAL menu.

   - Select CHANNEL 1 from SELECT GRAPHS menu.

   - Now you can select a portion of the force graph for averaging. Using the cursor keys, move the cursor to just before the impulse begins, where the force becomes nonzero. Press [ENTER].

   - Now move the cursor to the point at which the force was greatest. When the force was the greatest, the elastic cord was stretched to its maximum, and the velocity, v_f, was 0 m/s.

   - Press [ENTER], and read the value for the average force. Record the value as *average force, F*, in your data table.

   - Determine the length of the time interval by multiplying 0.02 s by the number of points used to determine average force. Record this product as *duration of impulse, Δt*, in your data table.

3. **Graph tracing** Return to the MAIN MENU. Select ANALYZE, VIEW GRAPH, and DISTANCE. Move the cursor along the distance-time graph to determine the stopping distance, *d*, which is the distance in meters from the point where the force was first exerted to the point where the force was maximum. Record this value as *stopping distance, d*, in your data table.

   If you have not yet completed the Procedure, return to step 12 in the Procedure now.

4. **Calculating** Because the final velocity is always 0 m/s, the change in velocity is equal to the initial velocity, with the sign reversed. For each trial, copy the value for the initial velocity to the change in velocity column in your data table, dropping the negative signs each time.

5. **Calculating** The impulse is the product of the average force and the length of time that force was applied, or *FΔt*. Use the average force and time interval data to determine the impulse for each trial. Record the impulse values in a new data table like the second one shown in the Data Tables section.

6. **Calculating** The change in momentum is equal to the mass of the cart times the change in velocity, or $m\Delta v$. Use the mass of the cart and the change in velocity to determine the change in momentum for each trial. Enter the values for change in momentum in your second data table, next to the impulse values.

CONCLUSIONS

7. **Evaluating results** Divide the difference between the impulse and the change in momentum by the average of the two, then multiply by 100 percent. Express the difference in the two values as a percentage. Record your results in your second data table.

8. **Evaluating models** Is the model consistent with the impulse-momentum theorem? Use your data to explain your answer.

9. **Reaching conclusions** Based on your results and your understanding of the impulse-momentum theorem, suggest a general rule that relates the average force exerted by a spring line and the stopping distance of a jet.

10. **Applying the model** Describe how the concepts of impulse and change in momentum apply to a jet that is landing on a short runway.

11. **Communicating results** How would you use your data when talking to new pilots to explain how spring lines work? Create an outline of key points that you would make in the seminar.

EXTENSIONS

1. **Extending research** Conduct library or Internet research to determine the mass of military jets, the lengths of runways on aircraft carriers, and the use of spring lines in landing.

2. **Designing presentations** Prepare a poster that explains the impulse-momentum theorem to jet pilots landing on the deck of aircraft carriers.

Technology Lab

Centripetal Acceleration

OBJECTIVES

- **Develop** a model to measure acceleration using a turntable.

- **Determine the relationship** between centripetal acceleration, radius, and angular velocity using the model.

- **Calculate** the radius and angular velocity to be used in new amusement park rides based on your data.

MATERIALS

- ✔ graphing calculator with link cable
- ✔ CBL system
- ✔ PHYSICS application loaded in calculator
- ✔ Vernier low-*g* accelerometer
- ✔ Vernier adapter cable
- ✔ turntable with three speed settings
- ✔ level
- ✔ meterstick
- ✔ masking tape
- ✔ stopwatch or watch with second hand
- ✔ 20 g slotted mass

Testing the Accelerations on Amusement Park Rides

You work as a safety inspector at a large amusement park. In addition to checking all the parts of the rides to make sure they are in good working order, you must also take measurements of speed, force, and acceleration to determine that each ride will work the way it is supposed to, and that riders will be able to safely enjoy the thrills of high accelerations and changes in speed. Many amusement park rides use rotational motion to create exciting effects for the riders.

Your amusement park has recently installed several rides that spin, rotate, and twirl. Your assignment is to inspect all the rides to make sure that they are designed properly and in good working order. In order to do this, you will first examine a model of rotational motion to determine the relationship between the angular velocity of a ride and the centripetal acceleration experienced by the riders. You will also examine how the centripetal acceleration varies with radius. These results will allow you to determine how the speed and size of different rides will affect the riders, and what safety measures must be taken to make each ride safe.

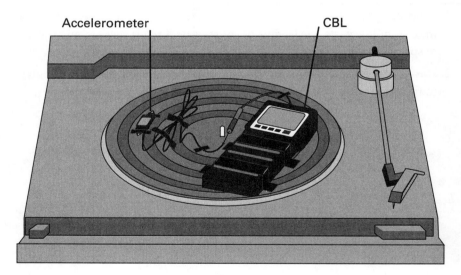

Accelerometer   CBL

SAFETY

- Perform this experiment in a clear area. Attach masses securely. Falling, dropped, or swinging masses can cause serious injury.

- Tie back long hair, secure loose clothing, and remove loose jewelry to keep them from getting caught in moving or rotating parts.

DEVELOPING THE MODEL

In this activity, you will use a turntable and the Vernier low-*g* accelerometer to confirm the mathematical relationship between angular velocity or tangential velocity and centripetal acceleration. When you have confirmed the relationship, you will use it to model the centripetal acceleration experienced by riders on amusement park rides.

Before you begin this activity, answer the following questions:

1. Place a 20-g mass on the rubber mat about 5 cm from the center of the turntable. Set the turntable to turn at 33⅓ rpm, and turn it on. Observe the motion of the mass. Is the speed of the mass constant or changing? Is the velocity of the mass constant or changing? Is the acceleration of the mass zero, constant and nonzero, or changing? If the acceleration is not zero, what is its direction?

2. Place a 20-g mass 5 cm from the center of the turntable. Set the turntable to turn at 33⅓ rpm, turn it on, and again observe the motion of the mass. After a few rotations, switch the speed to 45 rpm. Is the mass now undergoing less, the same, or more acceleration? Propose a mathematical relationship between centripetal acceleration and angular velocity.

3. Acceleration along a straight line can be either a positive or a negative quantity. Does speeding up while moving in the positive direction correspond to a positive or a negative acceleration? Does slowing down while moving in the positive direction correspond to a positive or a negative acceleration?

PROCEDURE

1. Connect the low-*g* accelerometer to CH 1 of the CBL system using the adapter cable.

2. Use the black link cable to connect the CBL unit to the calculator. Firmly press in the cable ends.

3. Turn on the CBL unit and the calculator. Start the PHYSICS application and proceed to the MAIN MENU.

4. Set up the calculator and CBL for the accelerometer.

   • Select SET UP PROBES from the MAIN MENU.

   • Select ONE as the number of probes.

   • Select ACCELEROMETER from the SELECT PROBE menu.

   • Confirm that the accelerometer is connected to CHANNEL 1, and press ENTER.

   • Select USE STORED from the CALIBRATION menu.

   • Select LOW-G from the ACCELEROMETER menu.

Part I

In this part, you will become familiar with the accelerometer and confirm the direction of acceleration relative to the arrow on the accelerometer.

5. Prepare to collect data with the accelerometer.

   • Select COLLECT DATA from the MAIN MENU.

   • Select TIME GRAPH from the DATA COLLECTION menu.

   • Enter "0.05" as the time between samples, in seconds.

- Enter "40" as the number of samples. The CBL will collect data for two seconds.
- Press ⟨ENTER⟩, then select USE TIME SETUP to continue. If you want to change the sample time or sample number, select MODIFY SETUP instead.

6. Rest the accelerometer on a smooth table top so that the arrow is horizontal and pointing to your right. Press ⟨ENTER⟩ to start data collection. Start with the accelerometer at rest. Without tilting it, move the accelerometer in the direction of the arrow for about 30 cm and stop. In other words, make the accelerometer start from rest, speed up and then slow down, finally stopping. All motion must be along the line of the arrow.

7. Press ⟨ENTER⟩ to view a graph of acceleration versus time. Since the accelerometer had to speed up in the direction of the arrow before later slowing down, is an acceleration in the direction of the arrow read as positive or negative? Record your answer in a data table like the one shown for Part I in the Data Tables section.

Part II

In this part, you will attach the accelerometer to the outer edge of the turntable platter, and you will measure centripetal acceleration at different angular velocities.

8. Place the turntable on a level surface. Check that the turntable platter is horizontal using the level. Securely tape the CBL to one side of the turntable platter. Securely tape the low-*g* accelerometer to the other side of the platter, near the outer edge of the platter, with the arrow pointing directly toward the spindle in the center. Also tape down the cable connecting the accelerometer and CBL. Make sure the CBL will not hit anything and the cable will not get caught on the needle arm or other obstacles as the turntable spins.

9. Measure the distance from the center of the accelerometer to the center of the turntable, and record the value in a data table like the one shown for Part II in the Data Tables section.

10. Next you will zero the sensor.
- Select ZERO PROBES from the MAIN MENU.
- Select CHANNEL 1 from the SELECT CHANNEL menu.
- With the turntable stationary, press on the CBL.

11. Set up the calculator and CBL for data collection.
- Select TRIGGERING from the MAIN MENU.
- Select MANUAL from the TRIGGERING menu.
- Select COLLECT DATA from the MAIN MENU.
- Select TIME GRAPH from the DATA COLLECTION menu.
- Enter "1" as the time between samples, in seconds.
- Enter "90" as the number of samples (the CBL will collect data for 90 seconds).

12. Press ⟨ENTER⟩, then select USE TIME SETUP to continue. If you want to change the sample time or sample number, select MODIFY SETUP instead.

13. You are now ready to collect centripetal acceleration data at three different angular velocities. For this lab, you will disconnect the CBL from the calculator while collecting data. After collecting data, you will reconnect the calculator and retrieve the data.

- Press [ENTER] and disconnect the black link cable from the CBL. The CBL will display "READY."
- Verify that the CBL, the low-*g* accelerometer, and the cable are taped down securely so that nothing will get hung up when the turntable spins.
- Set the turntable speed to 33⅓ rpm.
- Press [TRIGGER] on the CBL and wait 20 seconds.
- Turn on the turntable, letting it rotate at 33⅓ rpm.
- Wait 20 seconds and increase the speed to 45 rpm.
- Wait 20 seconds and increase the speed to 78 rpm.
- After 90 total seconds have elapsed, turn off the turntable.

14. Use the calculator to retrieve data from the CBL.

- Connect the black cable to the CBL, and press [ENTER] on the calculator.
- Select RETRIEVE DATA from the MAIN MENU.
- Press [ENTER] to retrieve the data from the CBL.
- Press [ENTER] to examine the graph.
- Press [ENTER] and select NO to return to the MAIN MENU.

15. Next, determine the average centripetal acceleration for each angular velocity.

- Select ANALYZE from the MAIN MENU.
- Select STATS/INTEGRAL from the ANALYZE MENU.
- Select STATISTICS from the STATS/INTEGRAL menu.
- Select CHANNEL 1 from the SELECT GRAPH menu.
- Select the portion of the acceleration graph when the turntable was rotating at 33⅓ rpm. Using the cursor keys, move the lower-bound cursor to the left side of the region. Press [ENTER].
- Now select the other boundary of the 33⅓-rpm region by moving the cursor to the right edge of the region (where the magnitude of the acceleration starts to increase as the angular velocity increases to 45 rpm). Press [ENTER].
- Press [ENTER], and read the value for the mean acceleration. Record the value, including the sign (+ or −), in your data table for Part II.
- Press [ENTER] to return to the MAIN MENU.

16. Repeat step 14 for 45- and 78-rpm portions of the graph. Record the values for mean acceleration in your data table for Part II.

Part III

In this part, you will measure how centripetal acceleration varies with radius by taking data at different radii while keeping the angular velocity constant at 78 rpm. Record which angular speed you will be using in the first line of a data table like the one shown for Part III in the Data Tables section.

17. The acceleration data at 78-rpm that you collected in Part II will serve as your first data point for this part. Copy the appropriate acceleration and radius values from Part II to the data table for Part III.

18. Move the accelerometer about 3 cm inward toward the center of the turntable. Tape it down securely with the arrow pointing directly toward the center of the turntable. Measure the distance from the center of the accelerometer to the center of the turntable and record the distance as a radius in your new data table.

19. Set the turntable speed to 78 rpm.

20. Prepare to collect acceleration data at various radii.

   - Select COLLECT DATA from the MAIN MENU.

   - Select TIME GRAPH from the DATA COLLECTION menu.

   - Enter "1" as the time between samples, in seconds.

   - Enter "30" as the number of samples (the CBL will collect data for 30 seconds).

   - Press ENTER, then select USE TIME SETUP to continue. If you want to change the sample time or sample number, select MODIFY SETUP instead.

21. Collect data with the calculator detached from the CBL again, as in step 13.

   - Press ENTER and disconnect the black link cable from the CBL. The CBL will display "READY".

   - Verify that the CBL, low-g accelerometer, and cable are taped down so that nothing will get hung up when the turntable spins.

   - Press TRIGGER on the CBL and wait five seconds.

   - Turn on the turntable.

   - Wait at least 25 seconds, then turn off the turntable.

22. Retrieve data from the CBL.

   - Connect the black cable to the CBL, and press ENTER on the calculator.

   - Select RETRIEVE DATA from the MAIN MENU.

   - Press ENTER to retrieve the data from the CBL.

   - Press ENTER to examine the graph.

   - Press ENTER and select NO to return to the MAIN MENU.

23. Next, determine the average centripetal acceleration.

   - Select ANALYZE from the MAIN MENU.

   - Select STATS/INTEGRAL from the ANALYZE MENU.

   - Select STATISTICS from the STATS/INTEGRAL menu.

   - Select CHANNEL 1 from the SELECT GRAPH menu.

   - Select a portion of the acceleration graph when the turntable was rotating at constant speed. Using the cursor keys, move the lower-bound cursor to the left side of the appropriate region. Press ENTER.

   - Now select the other edge by moving the upper-bound cursor to the right edge of the region. Press ENTER.

- Press ⌈ENTER⌉, and read the mean acceleration from the calculator. Record the value in your data table.

- Press ⌈ENTER⌉ to return to the MAIN MENU.

24. Move the accelerometer about 3 cm closer to the center of the turntable and tape it to the turntable with the arrow pointed directly toward the center. Repeat steps 20–23.

DATA TABLES

Part I

Acceleration relative to arrow on the accelerometer (positive or negative)	

Part II

	Radius (m)		
Angular speed (rpm)	Angular speed (rad/s)	(Angular speed)^2 (rad/s)^2	Centripetal acceleration (m/s^2)
33⅓			
45			
78			

Part III

Radius (m)	Centripetal acceleration (m/s^2)

ANALYSIS

1. **Converting units** Convert the angular speed values from the table for Part II from rpm (revolutions per minute) to radians per second. Remember that one revolution corresponds to 2π radians. Record the new values in your data table for Part II. Also calculate the angular speeds squared, and record the values in your data table for Part II.

2. **Graphing** Use your calculator or graph paper to plot a graph of your data from Part II with centripetal acceleration on the y-axis and angular speed squared on the x-axis. What can you tell about the relationship between centripetal acceleration and angular speed based on this graph?

3. **Interpreting graphs** As closely as you can, fit a straight line to your data points on the graph of centripetal acceleration versus angular speed squared. The line should also pass through the origin. What are the units of the slope of this line? What does the slope value represent?

4. Generating graphs Use your calculator or graph paper to plot a graph of data from Part III with centripetal acceleration on the y-axis and radius on the x-axis. What can you tell about the relationship between radius and centripetal acceleration based on this graph?

5. Interpreting graphs As closely as you can, fit a straight line to your data points on the graph of centripetal acceleration versus radius. The line should also pass through the origin. What are the units of the slope of this line? Note that rad/s and 1/s have the same dimensions, because a radian is dimensionless. Does this slope value correspond to any parameter in your experiment?

CONCLUSIONS

6. Interpreting data Based on the sign that the accelerometer gave when reading acceleration, is the centripetal force directed inward or outward?

7. Reaching conclusions From your graphs and your data, propose a mathematical relationship between centripetal acceleration, the square of angular velocity, and radius. Make sure the relationship is dimensionally consistent (i.e., that the units are equal on both sides of the equation).

8. Applying results One of the new rides is exactly like another ride that the park already has, except that the new ride has a radius of 1/2 the radius of the old ride. In order for the new ride to have the same centripetal acceleration as the old ride, what would the angular velocity have to be? If the new ride had the same angular velocity, what would the centripetal acceleration be?

9. Applying results A new ride called The Industrial Revolution has a radius of 6 m, and it spins at 21 rpm. Based on your data, what centripetal acceleration would a rider experience on this ride? What minimum force would the rider's seat have to be able to withstand for a 62 kg rider to remain securely seated throughout the ride? Remember, $F = ma$.

10. Applying results In one ride, riders stand inside a cylinder with their backs pressed against the wall. The cylinder begins to spin around the center axis. When the ride reaches a certain speed, the floor of the cylinder drops down and the riders stick to the wall. If the coefficient of static friction between the riders' clothes and the metal wall is 0.35, what is the minimum acceleration that will allow a 65 kg rider to stick to the wall? If the ride has a radius of 4.0 m, what angular velocity (in rpm) will it have to have?

EXTENSIONS

1. Making predictions A similar model can be used to investigate the actual motion of an object subjected to centripetal acceleration. Cover the turntable with a piece of clear acetate. Punch a hole so that it can fit over the spindle. Cover the acetate with a layer of fine dust and place an object that will slide, such as a penny, near the center. Turn the turntable on and allow the penny to slide off. Turn off the turntable, and the path of the penny will be left in the dust. Make predictions regarding the path the penny will take, then test your predictions.

2. **Making predictions** You can also use this setup to explore the motion of an object that is rolled across the turntable while it is turning. Make predictions regarding the path of a marble that is rolled from the outside edge of the turntable toward the center while it is turning. If you have time, test your predictions.

3. **Field investigations** Repeat this lab using a playground merry-go-round instead of a turntable. Before you begin, predict how changing the equipment may affect your results. Generate graphs of your data and compare them to the data you collected using the turntable. Note: Before performing this extension exercise, you should get permission and any special instructions from your teacher.

Newton's Law of Cooling

OBJECTIVES

- **Develop** a model that can be used to test the insulating ability of various fabrics.

- **Predict** temperature changes and energy loss over time using the model.

- **Derive** a cooling constant for an insulating fabric under wet or dry conditions.

- **Compare** the cooling constants for different fabrics to determine the best fabrics for use in cold or wet conditions.

MATERIALS

✔ graphing calculator with link cable

✔ CBL system

✔ PHYSICS application loaded in calculator

✔ TI or Vernier temperature probe

✔ 35-mm film canister with a hole in the lid

✔ samples of insulating fabric

✔ rubber bands

✔ hot water

✔ pipette

✔ balance

Outfitting for an Alpine Ski Trip

A friend of yours has decided to participate in an alpine cross-country ski trip. The trip involves camping in snow for three nights in cold, wet conditions. This is his first winter camping experience, and he wants to make sure that he takes the right clothing so that he remains safe and comfortable. When he asks an uncle for advice, his uncle insists that wool clothing is the best, especially in wet conditions. However, when he asks a salesperson at the local sporting goods store, she insists that new synthetic fleece materials are superior to wool under both wet and dry conditions.

Your friend has asked you to help him decide what kind of clothing to take on this trip. Your task is to test wool and synthetic fleece under both wet and dry conditions in the laboratory in order to determine the ability of each fabric to reduce the rate of heat loss.

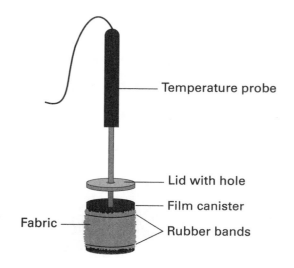

Temperature probe

Lid with hole

Film canister

Fabric

Rubber bands

SAFETY

- When using a burner or hot plate, always wear goggles and an apron to protect your eyes and clothing. Tie back long hair, secure loose clothing, and remove loose jewelry.

- Never leave a hot plate unattended while it is turned on.

- Use a hot mitt to handle equipment that may be hot. Allow all equipment to cool before storing it.

- If your clothing catches on fire, walk to the emergency lab shower if one is available or roll on the ground to put out the fire.

DEVELOPING THE MODEL

When you feel cold, it is often because energy is being transferred as heat away from your body into your surrounding environment. To measure the ability of different fabrics to slow the rate of energy loss, you will model a living body with a small quantity of hot water in a film canister. You will surround the canister with wool or synthetic fabric, either wet or dry.

Because the hot water in the film canister will be warmer than the surrounding air, energy will flow as heat from the water to the air in the room. However, it is difficult to measure this energy transfer directly. It is much easier to measure temperature changes over time. To relate temperature changes over time to energy transfer rate, you will use a mathematical model developed by Isaac Newton.

Newton's law of cooling states that the rate of energy transfer between a warm body and the surrounding environment is proportional to the difference in temperature, T_{diff}, between the body and its environment. This relationship can be expressed with the following equation:

$$\text{cooling rate} = -kT_{diff}$$

where k is a *cooling constant* for the given situation. The greater the value of the cooling constant for a given situation, the greater the rate of energy transfer.

The value of the cooling constant, k, can be calculated by measuring temperature differences over time according to the following exponential equation:

$$T_{diff} = T_0 e^{-kt}$$

where T_0 is the initial temperature difference (between the water and the air) and t is time (e is a constant, the base of the natural logarithm). By comparing the cooling constant, k, for different materials, you will be able to advise your friend on the best materials for the conditions.

Before starting this activity, answer the following questions:

1. According to the exponential equation $T_{diff} = T_0 e^{-kt}$, what is the value of T_{diff} when t is 0 (at the start of the experiment)?

2. According to the exponential equation $T_{diff} = T_0 e^{-kt}$, what happens to the value of T_{diff} when t becomes very large?

3. According to the equations above, what does a large cooling constant imply? Would the best insulator have a large or small value of k?

4. In a healthy living body, energy lost to the environment is replaced so that the temperature difference between the body and the environment remains constant. How does this make it difficult to determine the rate of energy transfer between the body and the environment?

PROCEDURE

1. Connect the temperature probe to the CH 1 input on the CBL. Use the black link cable to connect the CBL unit to the calculator. Firmly press in the cable ends.

2. Turn on the CBL unit and the calculator. Start the PHYSICS application and proceed to the MAIN MENU.

3. Set up the calculator and CBL for the temperature probe.

   • Select SET UP PROBES from the MAIN MENU.

   • Select ONE as the number of probes.

   • Select TEMPERATURE (for TI temperature probe or Vernier direct-connect temperature probe), VERN STD TEMP (for Vernier standard temperature probe) or VERN QIK TEMP (for Vernier quick-response temperature probe) from the SELECT PROBE menu.

- Confirm that the probe is attached to CHANNEL 1, and press [ENTER] to continue.

- Select USE STORED from the CALIBRATION menu. (Omit this step if using a TI probe.)

4. Determine room temperature. To do this, hold the probe in the air with nothing touching the probe tip.

- Select COLLECT DATA from the MAIN MENU.

- Select MONITOR INPUT from the DATA COLLECTION menu.

- Observe the temperature reading on the calculator. When it is stable, record the value in your data table as the room temperature.

- Press [+] to leave the monitor mode.

5. Choose a fabric sample or use the one assigned by your teacher. If necessary, cut the fabric in the shape of a rectangle so that you can wrap the fabric completely around the film canister without overlapping. If you are testing the fabric under wet conditions, completely wet the fabric with tap water at approximately room temperature, then wring out the fabric so that it is wet but not dripping. Use two rubber bands to fasten the fabric to your canister.

6. Obtain some hot water at about 60°C. You may be able to get water this hot from a hot water faucet. If necessary, use water that has been heated on a hot plate. Do not use water hotter than 60°C.

7. Place your canister on a balance. Zero the balance. Use a pipette to add 25 grams of hot water to the canister. Take the canister off the balance and place it back on your table.

8. Carefully push the temperature probe through the hole in the cap so that the end of the probe will be submerged in the water when the cap is on the canister. Place the lid containing the temperature probe onto the canister and press until the lid seals with a click. Make sure the end of the probe does not touch the bottom of the canister.

9. Set up the calculator and CBL for data collection.

- Select TIME GRAPH from the DATA COLLECTION menu.

- Enter "15" as the time between samples in seconds.

- Enter "80" as the number of samples.

- Press [ENTER], and select USE TIME SETUP to continue. If you want to change the sample time or sample number, select MODIFY SETUP instead.

- Select NON-LIVE DISPL from the TIME GRAPH menu.

10. Wait about 10 seconds for the temperature probe to reach the temperature of the water. Then collect your cooling data in the following way:

- Press [ENTER] to begin data collection. Data will be collected for 20 minutes.

- After the CBL shows DONE on its screen, the calculator will probably have turned itself off. Turn it back on and press [ENTER].

- Select RETRIEVE DATA from the MAIN MENU, and press [ENTER].

- Press [ENTER] to see your graph. Sketch the graph for later reference.

- Press [ENTER] and select NO to return to the MAIN MENU.

DATA TABLES

Fabric type	
Wet or dry?	
Room temperature (°C)	

A	
B	
Initial temperature difference, T_0	
Cooling constant, k	

Fabric and conditions of test	Cooling constant, k

ANALYSIS

1. **Calculating** The PHYSICS application stores temperature data in the calculator list L_2. Because Newton's law of cooling uses the difference between the sample temperature and room temperature, you must subtract the room temperature from the measured absolute temperatures to get the temperature difference, T_{diff}. The following steps convert the absolute temperature values in L_2 to the required temperature differences:

   - Select QUIT from the MAIN MENU.
   - Press "[2nd] L2 [–] (*room temperature*) [STO▶] [2nd] L2 " where (*room temperature*) is the temperature you determined in step 4 of the Procedure.
   - Restart the PHYSICS application and proceed to the MAIN MENU.

2. **Curve fitting** Use the following steps to fit an exponential function of the form $y = Ae^{-Bx}$ to your temperature difference versus time data.

   - Select ANALYZE from the MAIN MENU.
   - Select CURVE FIT from the ANALYZE MENU.
   - Select EXPONENT L_1, L_2 from the CURVE FIT menu.
   - Record the curve fit parameters A and B in your data table.
   - Press [ENTER] to see a graph of your data with the fitted function.

3. **Applying the model** As shown above, Newton's law of cooling can be stated with the following exponential equation:

$$T_{diff} = T_0 \, e^{-kt}$$

   Your graph now shows temperature difference, T_{diff}, versus time, t. You have also used the calculator to fit the following function to your data:

$$y = Ae^{-Bx}$$

   In this generic equation, y corresponds to the temperature difference, T_{diff}, x corresponds to the time, t, A corresponds to the initial temperature, T_0, and B corresponds to the cooling constant, k. Using these relationships, record the values of T_0 and k in your data table.

CONCLUSIONS

1. **Evaluating models** Study the graph of the fitted curve. Does the exponential equation for Newton's law of cooling fit your data well?

2. **Interpreting graphs** What does the graph indicate about the rate of energy transfer as the temperature of the water gets closer to the air temperature?

3. **Making predictions** Write out the full exponential equation for T_{diff} using the value for k that you found in the experiment. Use the equation to calculate what the temperature of the water would be after 60 minutes of cooling. Then calculate what the temperature of the water would be after 10 hours of cooling. According the model, will the water in the canister ever reach the same temperature as the surrounding air? Explain.

4. **Thinking critically** If you had started this experiment with a smaller or larger temperature difference, would the value you reported for k be different? One way to approach this question is to imagine that you started collecting your data after the temperature had already dropped several degrees. In this case, would the rest of the graph look the same?

5. **Presenting data** Write the value for your cooling constant, k, on the chalkboard. Also write the type of fabric you tested and whether the fabric was wet or dry. When each group has presented its data, write the values in a data table like the last table in the Data Tables section. If more than one group tested the same type of fabric under the same conditions, average the values for k to get a single value for that fabric type and condition.

6. **Reaching conclusions** Based on the results from the whole class, what type of fabric would you recommend that your friend take on his alpine skiing and camping trip?

EXTENSIONS

1. **Applying the model** Based on what you have learned in this lab, explain why it takes fewer BTU's of energy to keep a house at 65° F than it does to keep the same house at 68° F when the temperature outside is 40°.

2. **Applying the model** A coffee drinker is faced with the following dilemma. She is not going to drink her hot coffee with cream for ten minutes, but wants it to still be as hot as possible. Is it better to immediately add the room-temperature cream, stir the coffee, and let it sit for ten minutes, or is it better to let the coffee sit for ten minutes and then add and stir in the cream? Which results in a higher temperature after ten minutes? If you have time, use your temperature probe and water at different temperatures to test your hypothesis. Explain your results in terms of Newton's law of cooling.

3. **Designing experiments** Mathematical models similar to the one for the cooling of a liquid can be applied to other phenomena in nature. For example, radioactivity and RC circuits can both be modeled by exponential functions of the form $y = Ae^{-B*x}$. Research other phenomena that are modeled by exponential functions. Design an experiment to test the exponential nature of one of these phenomena in your physics lab. Note: Do not perform your experiment unless you are instructed to do so by your teacher.

HOLT PHYSICS
Technology Lab

Pendulum Periods

Designing a Clock Pendulum

The owner of an antique store has just purchased an old grandfather clock that appears to be in good working order. She was able to purchase the clock for a fraction of its potential value because the original pendulum was missing. To restore the clock to working order, the owner wants you to design a pendulum for the clock that will allow it to keep good time. By studying the gears, she has determined that the original pendulum on this clock had a period of 1.875 s. She has asked that you provide her with the following pendulum specifications: length of pendulum, mass of pendulum bob, and angle of swing.

Your first task is to design three separate experiments to test the effects of amplitude (angle of swing), pendulum length, and mass on the period of a clock pendulum. Once you have collected your data, you will be asked to provide her with the relevant information to build a pendulum for the clock.

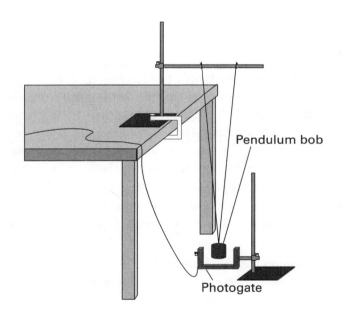

Pendulum bob

Photogate

OBJECTIVES

- **Design** experiments to determine the factors that may affect the period of a pendulum.

- **Test** the relationship between length, mass, amplitude, and the period of a pendulum.

- **Recommend** the specific characteristics of a clock pendulum using your data.

MATERIALS

- ✔ graphing calculator with link cable
- ✔ CBL system
- ✔ PHYSICS application loaded in calculator
- ✔ Vernier photogate and CBL adapter cable
- ✔ 2 support stands and clamps
- ✔ table clamp or 1 kg mass
- ✔ masses of 100, 200, and 300 g with hooks
- ✔ meterstick
- ✔ protractor
- ✔ graph paper

SAFETY

- Perform this experiment in a clear area. Attach masses securely. Falling, dropped, or swinging masses can cause serious injury.

- Tie back long hair, secure loose clothing, and remove loose jewelry to keep them from getting caught in moving or rotating parts.

DEVELOPING THE MODEL

To design a clock pendulum, you first need to conduct controlled experiments to test the dependence of a pendulum's period on length, mass, and amplitude of swing. In your experiments, a string with a mass attached will simulate the clock pendulum. You will use a photogate to measure the period of the pendulum. As the pendulum moves back and forth, it will pass through the photogate, and the CBL will measure the time elapsed between successive passes.

Before beginning this experiment, answer the following questions:

1. For each trial in your experiments, you should measure five complete periods. The first pass through the photogate starts the timer. How many times will the pendulum pass through the photogate for each trial?

2. The PHYSICS program will automatically calculate the pendulum period once the data are collected. Why does the photogate calculate the period based on the time elapsed between every other pass rather than every pass through the photogate?

4. Make a simple pendulum by tying a 100 cm string to a mass. Tie the string to a support rod and let the mass swing. While it is swinging, predict what might happen to the period if the length of the string were changed, and write down your prediction.

5. Swing the pendulum again, and make similar predictions regarding changes in mass or amplitude. Write down your predictions.

PROCEDURE

1. Place a support stand on the edge of a table and secure it to the table with a clamp or by placing a heavy mass on the base. Tie two strings of equal length to the 200 g mass. Attach the strings about 15 cm apart to a horizontal rod extended from the support stand. The pendulum should hang over the edge of the table, as shown on the previous page. This arrangement will let the mass swing only along a line, and will prevent the mass from striking the photogate. The length of the pendulum is the distance from the point on the rod halfway between the strings to the center of the mass. The pendulum length should be at least 1 m (you do not need an exact measurement now).

2. Attach the photogate to a second support stand. Place the support stand on the floor and position the photogate so that the mass blocks the photogate while hanging straight down. Connect the photogate to the CH 1 port on the CBL. Use the black link cable to connect the CBL to the calculator. Firmly press in the cable ends.

3. Turn on the CBL and the calculator. Start the PHYSICS application and proceed to the MAIN MENU.

4. Set up the calculator and CBL for the photogate.
   - Select SET UP PROBES from the MAIN MENU.
   - Select ONE as the number of probes.
   - Select PHOTOGATE from the SELECT PROBE menu.
   - Proceed to the TIMING MODES menu.
   - Select CHECK GATE to see that the photogate is functioning.

5. Observe the reading on the calculator screen. Temporarily hold the mass out of the center of the photogate. Block the photogate with your hand; note that the photogate is shown as blocked. When you remove your hand, the display should change to unblocked. Press ⌊ + ⌋ to return to the TIMING MODES menu.

   • Select PENDULUM from the TIMING MODES menu.

   • Enter "5" for the number of oscillations.

6. Temporarily hold the mass out of the beam of the photogate. Press ⌊ENTER⌋ to prepare the photogate.

7. Now you can perform a trial measurement of the period of your pendulum. Hold the mass displaced about 10° from vertical and release. (For a pendulum that is 100 cm long, that corresponds to pulling the mass about 15 cm to the side.) After the mass has passed through the photogate for five complete cycles (eleven passes through the photogate), the average period is displayed on the calculator screen.

8. To measure another period, press ⌊ENTER⌋ and select YES; again enter "5" for the number of oscillations. When you are ready, press ⌊ENTER⌋ to prepare the photogate. You will use this method for each period measurement below.

Part I Amplitude

9. Measure the period for five different amplitudes. Choose a wide range of amplitudes, from just barely enough to unblock the photogate, to about 30°. For each trial, use the protractor to measure the amplitude before releasing the mass. Repeat step 8 to collect the data and record the data in a data table like the one shown for Part I in the Data Tables section.

Part II Length

10. Now you will measure the effect of changing pendulum length on the period. Use the 200 g mass and a consistent amplitude of 10° for each trial. Vary the pendulum length in steps of 10 cm, from 100 cm down to 50 cm. Be sure to measure the pendulum length from the rod to the *center* of the mass. You may have to raise the photogate by placing a stack of books under the lower support stand. Repeat step 8 for each length. Record your data in a data table like the one shown for Part II in the Data Tables section.

Part III Mass

11. Use three different masses to determine if the period is affected by changing the mass. For each mass trial, keep the length of the pendulum the same and use a constant amplitude of 10°. Repeat step 8 for each trial. Record your data in a data table like the one shown for Part III in the Data Tables section.

DATA TABLES

Part I Amplitude

Amplitude (°)	Average period, T (s)

Part II Length

Length, l (cm)	Average period, T (s)

Part III Mass

Mass, m (g)	Average period, T (s)

ANALYSIS

1. **Graphing data** Using your graphing calculator or graph paper, plot a graph of pendulum period versus amplitude in degrees. Scale each axis from the origin. According to your data, does the period appear to depend on amplitude? Explain.

2. **Graphing data** Using your graphing calculator or graph paper, plot a graph of pendulum period versus length. Scale each axis from the origin. According to your data, does the period appear to depend on length? Explain.

3. **Graphing data** Using your graphing calculator or graph paper, plot a graph of pendulum period versus mass. Scale each axis from the origin. According to your data, does the period appear to depend on mass? Do you have enough data to answer this conclusively? Explain.

4. **Curve fitting** To examine more carefully how the period depends on the pendulum length, create the following two additional graphs of the same data: T versus l^2 and T versus *the square root of l*. Of the three period–length graphs, which plot is most nearly a straight line that goes through the origin?

5. **Interpreting graphs** Using the graph you chose in the last step, write a simple equation containing a proportionality constant that shows the relationship between T and l. Use your data to determine the value of the proportionality constant. (Hint: the proportionality constant will equal the slope of the line on the graph you chose.)

CONCLUSIONS

6. **Reaching conclusions** Use the equation you wrote to determine the appropriate length of the pendulum for the antique dealer's clock. Recall that the desired period is 1.875 s.

7. **Designing systems** If you were to design a pendulum clock, what would be the simplest way to make the clock adjustable to run faster or slower. Explain why you think so.

8. **Thinking critically** In designing real clock pendulums, the length of a pendulum is often stated in terms of the distance between its center of mass and the pivot point, rather than simply the length of the pendulum. Why might this be an important distinction?

9. **Presenting results** Write a short description of the pendulum for the antique dealer that contains all of the essential information, including a recommendation of the length, mass, and targeted amplitude of the pendulum. Finally, make design suggestions so that the pendulum period is adjustable.

EXTENSIONS

1. How could temperature affect a pendulum clock? What other environmental factors might cause a clock to run faster or slower?

2. Try conducting the experiment with a larger range of amplitudes. Use the library or other reference sources to find a mathematical expression that models this relationship and check to see if your data fit the model.

3. Use the graph you chose and your data to determine a local value for the free-fall acceleration, g.

4. Use an air table and air table puck to model simple harmonic motion. Tip the air table to a variety of angles and determine the relationship between the period of oscillation and the angle.

Sound Waves and Beats

OBJECTIVES

- **Develop** a setup that allows you to analyze sound waves for frequency and amplitude.

- **Analyze** the waveforms of a tuning fork and a musical instrument.

- **Model** waveforms using a mathematical expression derived from your data.

- **Observe** beat frequencies when two waveforms are combined.

- **Evaluate** your ability to tune by ear based on beat frequencies in the combined waveform.

MATERIALS

✔ graphing calculator with link cable

✔ CBL system

✔ PHYSICS application loaded in calculator

✔ Vernier microphone

✔ tuning fork

✔ rubber striker for tuning fork

✔ stringed instrument (or other musical instrument that may be tuned)

Tuning a Musical Instrument

Perfect pitch, also called *absolute pitch,* is the ability to identify a note on the musical scale without any reference note. A person with absolute pitch can be blindfolded, hear a D on the piano, and correctly identify the note. Very few people possess absolute pitch, but many people possess *relative pitch*. If a person with relative pitch is given a known reference note on the piano, they can use that tone to identify pitches relative to that reference note. A third group of people can only tell if tone is higher or lower in pitch than another tone. Finally, a fourth group is described as *tone-deaf*. Tone-deaf individuals are not able to tell if one tone is higher than another unless the two pitches are several notes or more apart.

How good are your ears? In this activity, you will test your ability to match two tones using your own perception of pitch differences. Your task is to tune one string of a stringed instrument by comparing the pitch produced by the string to the pitch produced by a tuning fork. When you think you have tuned the instrument correctly, you will evaluate your perception of pitch differences by using a CBL, a microphone, and a graphing calculator to analyze the tones and look for beat frequencies.

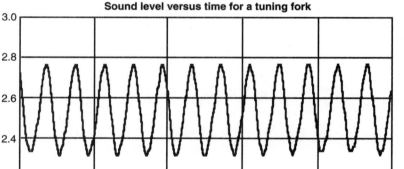

Figure 13-1

SAFETY

- Review lab safety guidelines at the front of this booklet. Always follow correct procedures in the lab.

- Please try to minimize excess noise during this lab. You and other students will need to be able to hear subtle differences in pitch. Also, background noise can corrupt sound data.

DEVELOPING THE MODEL

The microphone and CBL system can detect sound waves, analyze them, and present the data in the form of a graphical waveform, like the graph in **Figure 13-1** on the previous page. Because the sound waves are measured electronically, the sound level is shown in units of volts (V). By using the data to create a mathematical model, you will be able to compare the amplitude and frequency of a tuning fork and a musical instrument. Answer the following questions before beginning this activity:

1. Sound waves can be described as a series of compressions and rarefactions that occur in a medium. Sketch the graph shown in **Figure 13-1** and label a point at which the air molecules carrying a sound are at maximum compression. Also label a point where the air molecules are at a maximum rarefaction.

2. One property of sound is its volume. How is the amplitude of the graphical model of sound shown in **Figure 13-1** related to the volume of the sound?

3. The period of a wave is the time that it takes for one complete cycle of a wave to elapse. Determine the period for the sound waves represented in **Figure 13-1**. To determine the period, find the elapsed time between the first and last peak shown on the graph and divide by the number of cycles.

4. How is the frequency of a sound, f, related to the period, T, of the sound waves? Use this relationship to calculate the frequency of the sound represented in **Figure 13-1.**

5. The displacement of the particles in air carrying a sound wave can be modeled with the following sinusoidal function:

$$y = A \sin(2\pi f t)$$

where y refers to the change in air pressure that makes up the wave, A is the amplitude of the wave, f is the frequency, and t is time. Describe how the graph will change when the value of A goes up or down. Describe what happens to the graph when the value of f increases or decreases.

PROCEDURE

1. If your calculator is in degree mode, change it to radian mode.

2. Connect a Vernier microphone to the CH1 input on the CBL. Use the black link cable to connect the CBL unit to the calculator. Firmly press in the cable ends.

3. Turn on the calculator and CBL unit. Start the PHYSICS application and proceed to the MAIN MENU.

4. Set up the calculator and CBL for the microphone.
   - Select SET UP PROBES from the MAIN MENU.
   - Select ONE as the number of probes.
   - Select MICROPHONE from the SELECT PROBE menu.
   - Press ENTER to continue to the SELECT MICROPHONE menu.
   - Select CBL for the type of microphone.
   - Select WAVEFORM from the COLLECTION MODE menu.

Part I The Waveform of a Tuning Fork

5. Strike the tuning fork against the rubber striker, hold the tuning fork close to the microphone, and press [ENTER]. You should see a sinusoidal waveform on the graphing calculator, similar to the one in **Figure 13-1.** Be sure to strike the tuning fork against a soft object such as a rubber tuning fork striker, a rubber stopper, or the rubber sole of a shoe. Striking a tuning fork against a hard object can damage the fork and will also cause the sound to have unwanted harmonics.

6. If you strike the tuning fork too hard or too softly or if there is excessive background noise, the waveform may be very rough; if so, strike the fork and collect data again. When you have a good, clean sinusoidal waveform, make a sketch of the graph for later reference.

7. Using the cursor keys, trace across the graph. Record the times for the first and last peaks of the waveform. In a data table like the first one shown in the Data Tables section, record the number of complete cycles that occur between your first measured time and the last. Divide the difference, Δt, by the number of cycles to determine the period of the tuning fork. Record Δt and the period in your data table.

8. Trace across the graph again, and find the maximum and minimum y values for an adjacent peak and trough. Record these two values in a data table like the second one shown in the Data Tables section. Note that the units of y-values are volts (V), because the CBL determines signal strength from the microphone using electric potential difference, or voltage.

9. Temporarily leave the PHYSICS application. Press [ENTER], select NO, select RETURN TO MAIN, and select QUIT.

10. Calculate the amplitude of the wave by taking half of the difference between the maximum and minimum y values. Record the amplitude in volts in your second data table.

11. Use the period you calculated in step 7 to calculate the frequency of the tuning fork in Hz. Record the frequency in your first data table.

12. To create a mathematical model of your data, you will now enter the equation, A*sin(B*X+C), into your calculator. In the model, A corresponds to the amplitude, A, B corresponds to $2\pi f$, and X is the time, t, in seconds. C is an adjustment factor that shifts the model to the left or right.
    - From the calculator home screen, press [Y=].
    - Press [CLEAR] to clear the Y_1 equation.
    - Enter "A*sin(B*X + C)" in the Y_1 line.

13. To see how well the model fits the data, you need to enter values for the parameters A, B, and C of the model equation and plot it with the data. To enter values in your model,
    - Start the PHYSICS application and proceed to the MAIN MENU.
    - Select ANALYZE from the MAIN MENU.
    - Select ADD MODEL from the ANALYZE MENU.

- Select ADJUST A from the MODEL MENU.

- Enter an estimate for the value of A based on your data, ending with [ENTER]. Your estimate should be the amplitude you calculated from the data.

- Repeat for B. Your estimate for B should be 2π times the frequency you calculated from the data.

- Repeat also for C. A positive value of C will shift the model to the left, and a negative value of C will shift the model to the right.

The calculator will plot the data and your model with the current values of A, B, and C. Use the values of amplitude and frequency you derived from your data to begin with. Adjust each parameter until you see what each one does to the graph and until you have a good match of your model to the data.

14. When you have found a good match between your mathematical model and your tuning fork data, record the final values of A and B in a data table like the third one shown in the Data Tables section. Press [ENTER] and then MODEL OFF+RTRN to go back to the MAIN MENU.

Part II The Waveform of a Musical Instrument

15. Use the tuning fork to tune one of the strings on your instrument. Choose a string that has a pitch close to that of the tuning fork when the string is "open" (no fingers pressing on the string). Adjust the pitch of the string slowly and carefully by turning the appropriate tuning peg until you think that the string and the tuning fork are at the same pitch. Do not overtighten the string.

16. Now generate a graph for the sound produced by the string.

- Select COLLECT DATA from the MAIN MENU.

- Repeat step 5 for the string you just tuned. Hold the string that was tuned very near the microphone. When plucking the string, try to produce a volume that is similar to the volume produced by the tuning fork.

- Because a stringed instrument naturally produces more harmonics than a tuning fork, the waveform will not be as clean. However, you may want to collect data a few times to get the best possible data.

- Repeat steps 6–11 for the tuned string data.

17. Now you are going to fit the mathematical model to the data for the tuned string. Repeat step 13, selecting new parameters A, B, and C to fit the model to the data.

18. When you have found a good match between your mathematical model and the data for the stringed instrument, record the final values of A, B, and C in your third data table. Press [ENTER] and then MODEL OFF+RTRN to go back to the MAIN MENU.

DATA TABLES

Procedure

Source of sound	Number of cycles	First maximum (s)	Last maximum (s)	Δt (s)	Period (s)	Calculated frequency (Hz)
tuning fork						
tuned string						

Source of sound	Peak (V)	Trough (V)	Amplitude (V)
tuning fork			
tuned string			

Source of sound	Amplitude parameter, A (V)	Frequency parameter, B (s^{-1})	Horizontal parameter, C	$f = B / 2\pi$ (Hz)
tuning fork				
tuned string				

Analysis

Number of cycles	First maximum (s)	Last maximum (s)	Δt (s)	Beat period (s)	Beat frequency (Hz)

ANALYSIS

Now you will examine the two model waveforms when they are combined; you will look in particular for the presence of beats. The presence or absence of beats will help you decide how well you tuned your instrument.

1. **Programming the calculator** Return to the calculator home screen. Now create a new mathematical model that is the sum of the two individual models.

   - From the calculator home screen, press $\boxed{\text{Y=}}$.
   - Press $\boxed{\text{CLEAR}}$ to clear the Y_1 equation.
   - In the Y_1 line, enter A*sin(B*X+C), but this time substitute in the actual numerical values of the parameters A, B, and C, using the data for the tuning fork in your data table.
   - In the Y_2 line, enter A*sin(B*X+C), substituting in the numerical values of the parameters A, B, and C, using the data for the tuned instrument in your data table.
   - Move the cursor down to the Y_3 line and enter $Y_1 + Y_2$ in the Y_3 line.
   - Press $\boxed{\text{2nd}}$ STAT PLOT.
   - Turn off all stat plots.

2. **Programming the calculator** Before viewing the combined model graph, you will need to resize your window.

- Press the [WINDOW] button on the calculator.
- For Xmin, type 0.
- For Xmax start with 0.5. This will allow you to see beats that are occurring in the first 0.5 seconds. You may need to make the window larger or smaller depending upon how closely you tuned your instrument to the tuning fork. If your two values of B are very different, you may need to decrease the value of Xmax. If the values were very similar, you may need to increase Xmax to 1.0 second or more.
- For Xscl, type 0.1
- For Ymin, choose a negative value whose absolute value is a little larger than the largest sum of parameters A and C. This will allow you to see the whole wave when the two waves are added together.
- For Ymax, choose the corresponding positive value.
- For Yscl, choose 0.1.

3. **Generating graphs** Press the [GRAPH] button to view the graph of the two waves added together. If necessary, increase or decrease Ymax and Ymin to compress or stretch out the amplitude to fill, but remain within, the display. Increase or decrease Xmax until four to six cyclical variations in amplitude can be seen in the display.

4. **Graph tracing** Using the cursor keys, determine the times for the first and last amplitude maxima. Divide the difference between these times, Δt, by the number of cycles to determine the period of beats in seconds. Calculate the *beat frequency* in Hz from the beat period. Record these values in a data table like the fourth one shown in the Data Tables section.

5. **Interpreting graphs** Describe the appearance of the final combined model. At points in time when the amplitude is at a maximum, what must be true about the amplitudes of the individual waves at these times? When the amplitude of the combined model is zero, or at a minimum, what must be true about the amplitudes of the individual waves?

CONCLUSIONS

6. **Applying the model** The model parameter B corresponds to $2\pi f$. Use your fitted model to determine the frequency for both the tuning fork and the tuned string ($f = B/2\pi$). Enter the values in your third data table. Compare these frequencies to the frequencies calculated earlier. Which would you expect to be more accurate? Why?

7. **Evaluating models** Did your mathematical model fit the waveform data for the tuning fork and the stringed instrument well? In what ways was the model similar to the data and in what ways was it different?

8. **Interpreting data** How many beats per second were observed in the combined model? How does this compare to the difference in frequencies of the two individual waveforms?

9. **Making predictions** How many beats would have been observed if the string had been tuned to exactly the same pitch as the tuning fork? Explain.

10. **Reaching conclusions** Evaluate your ability to tune by ear based on the number of beats you observed.

11. **Applying results** Many musicians who play stringed instruments can't tell the difference in pitch between two tones played separately that are less than 5 Hz apart in frequency. However, they are able to tune their instruments to within 1 Hz of the proper relative pitch using only their ears. How are they able to accomplish this?

EXTENSIONS

1. **Extending the model** Use the microphone to examine the pattern you get when you play two adjacent notes on an electronic keyboard. When doing this, you should choose VOLTAGE for the probe so that you may set the data collection rate. Under COLLECT DATA, select TIME GRAPH. Enter "0.03" as the time between samples and "99" as the number of samples. Explore how the beat frequencies change as the two notes played get further and further apart.

2. **Extending research** Commercial products called *active noise cancellers* use headphones, microphones, and some electronics to reduce noise in noisy environments where the user must still be able to hear (for example, on airfields). Conduct library or Internet research to find out how these products work.

HOLT PHYSICS
Technology Lab

Polarization of Light

Testing Polarized Sunglasses

A local consumer research company is preparing to write a report on sunglasses. The report will cover many aspects of the performance of sunglasses, including durability, fit, comfort, and the ability to reduce glare. The company has asked for your help in addressing one question in particular: Is the reduction in glare that polarized lenses provides worth the higher price that consumers must pay?

The most common source of glare is light reflected off horizontal surfaces, such as a road or the surface of water. Light reflected off horizontal surfaces becomes polarized along the horizontal axis. Polarized sunglasses contain lenses that are polarized vertically so that horizontally polarized light cannot pass through.

Your task is to study the ability of polarized lenses to block polarized light. You will first use two high-quality polarizing filters and an artificial light, then you will use a pair of polarized sunglasses and sunlight reflected off a horizontal surface. You will compare the results you get in these two cases and you will write a short consumer report based on your data.

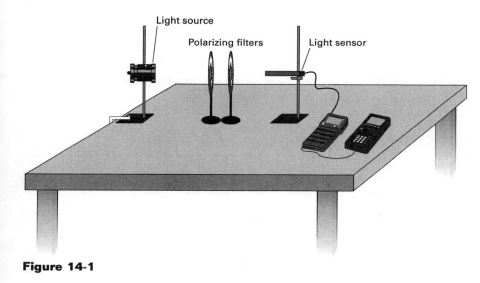

Light source

Polarizing filters   Light sensor

Figure 14-1

OBJECTIVES

- **Develop** an experiment that tests the ability of polarized sunglasses to reduce glare.

- **Measure** the transmission of light through two polarizing filters as a function of the angle between their axes.

- **Measure** the transmission of reflected light (glare) through a pair of polarized sunglasses as a function of angle.

- **Evaluate** the effectiveness of polarized lenses in a consumer report.

MATERIALS

- ✔ graphing calculator with link cable
- ✔ CBL system
- ✔ PHYSICS application loaded in calculator
- ✔ TI or Vernier light sensor
- ✔ DC light source
- ✔ two polarizing filters with marked axes
- ✔ pair of polarized sunglasses
- ✔ smooth horizontal surface to reflect light
- ✔ cardboard screen
- ✔ three support stands with clamps
- ✔ tape
- ✔ toothpick
- ✔ protractor

SAFETY

- Review lab safety guidelines at the front of this booklet. Always follow correct procedures in the lab.

- Avoid looking directly at a light source. Looking directly at a light source may cause permanent eye damage.

DEVELOPING THE MODEL

To evaluate polarized sunglasses for their ability to reduce glare, you will first observe how the transmission of light can be reduced using two polarizing filters. You will use a light sensor to measure the amount of light that passes through both of the filters when they are at different angles. You will generate a graph of your data that shows how the transmission of light changes as the angle between the axes of the filters changes.

Then you will perform a similar test using reflected sunlight and a pair of polarized sunglasses. After you have collected your data, you will be able to compare light reduction under ideal conditions with the reduction of glare using polarized sunglasses.

Answer the following questions before beginning this activity.

1. Why would it be important to use reflected glare as a source of polarized light rather than light that has passed through a polarized filter when testing the ability of sunglasses to reduce glare?

2. In the 1800's, Malus proposed a mathematical model to predict the intensity, I, of polarized light transmitted through a polarizing filter or lens. The relationship is

$$I = I_0 \cos^2 \theta$$

where I_0 is the intensity when the angle θ between the axes of the polarizers is zero. When the difference between the two axes of polarization is 0°, what is the value of $\cos^2 \theta$?

3. When the difference between the axes of polarization is 90°, what is the value of $\cos^2 \theta$?

4. Sketch a graph of I versus θ based on your answers to questions 2 and 3.

5. Two strategies to reduce the amount of light entering your eyes are to place darker lenses into a pair of sunglasses or to place polarized lenses into the sunglasses. Describe how each of these will affect what you actually experience.

PROCEDURE

Part I Two Polarizing Filters

1. If your calculator is in radian mode, change it to degree mode.

2. Mount the light source, polarizing filters, and light sensor on stands, and position them so that light passes through both filters and then into the sensor. You will rotate only one filter to change the transmission; the other filter, the light source, and the sensor must not move. Turn on the light source.

3. Attach the light sensor to the CH 1 input on the CBL. Use the black link cable to connect the CBL unit to the calculator. Firmly press in the cable ends.

4. Turn on the CBL unit and the calculator. Start the PHYSICS application and proceed to the MAIN MENU.

5. Set up the calculator and CBL for the light sensor.

   • Select SET UP PROBES from the MAIN MENU.

   • Select ONE as the number of probes.

   • Select LIGHT from the SELECT PROBE menu.

- Confirm that the light sensor is connected to CHANNEL 1, and press ENTER.
- Select COLLECT DATA from the MAIN MENU.
- Select MONITOR INPUT from the DATA COLLECTION menu.

6. Rotate the analyzer until the reading on the calculator is maximized. This is your zero angle transmission. The axis marks on the two filters should be parallel, with a 0° difference.

7. Set the filters so their axes are at right angles (e.g., one at 0°, one at 90°). Read the current light level, which is effectively the background light level:

- Note the reading on the calculator screen. It should be stable. If not, you may have excessive room light. Reduce room light until the reading is stable.
- Record the background light level in a data table like the one shown in the Data Table section.
- Press + to return to the DATA COLLECTION menu.

8. Set up the calculator and CBL for data collection.

- Select TRIGGER/PROMPT from the DATA COLLECTION menu.

In this mode, light intensity will only be measured when the TRIGGER button is pressed. You will then type the analyzer angle in degrees and press ENTER to complete the data point.

9. Take a data point:

- Rotate the analyzer to the 0° position.
- Press TRIGGER to take the first point and enter "0" for the angle.
- Press ENTER to complete the entry.
- Select MORE DATA from the DATA COLLECTION menu to continue.

10. Rotate the analyzer by 15°, press TRIGGER and enter "15" for the angle. Repeat this process, entering "30" for the next angle, and so forth, until you have rotated the analyzer through one full revolution, or 360°. After the last point, select STOP AND GRAPH to end data collection.

11. You should now see a graph of light intensity versus the analyzer angle. Trace across the graph with the cursor keys. Note the maximum value of the intensity and record it in the data table.

12. To compare Malus's model, $I = I_0\cos^2\theta$, with your data, you need to enter the equation "A*(cos(X))^2 + B" in your calculator. The variable A will correspond to the maximum intensity I_0. The B term represents the additional intensity caused by background light. To enter the equation in your calculator, you need to temporarily leave the PHYSICS application. Press ENTER, select NO, and select QUIT. Then:

- From the calculator home screen, press Y=.
- Press CLEAR to clear the Y_1 equation.
- Enter "A*(cos(X))^2 + B" in the Y_1 line.
- Start the PHYSICS application again and proceed to the MAIN MENU.

13. To plot your data and Malus's model on the graph together:

- Select ANALYZE from the MAIN MENU.
- Select ADD MODEL from the ANALYZE MENU.

- Select ADJUST B.

- Enter the background intensity you observed.

- The graph you see will not yet show the model, since you have not set the constant A. Press [ENTER] to prepare to set A.

- Select ADJUST A.

- Enter the difference between the maximum intensity from step 2 and the background intensity.

- Press [ENTER] to view a graph with the model superimposed on your data.

- Sketch the graph for later reference.

Part II Testing Sunglasses

14. Use a table in direct sunlight near a window. Lower the window shade part way or position a piece of cardboard to partially block the sunlight, as shown in **Figure 14-2.** Clamp the light sensor to a support stand behind the cardboard so no direct sunlight enters the sensor, only reflected light. Aim the sensor so that it points directly at the reflecting surface of the table. Make sure the CBL, the calculator, and the sensor cable are out of the way.
Note: If the sun is not out or there is no table in direct sunlight, you may use the light source from Part I instead of sunlight.

15. Because the axis of the reflected light is horizontal, the axis of polarization for the sunglasses should be vertical. Use tape to attach a toothpick to one of the lenses so that the toothpick is oriented vertically when the sunglasses are in a normal upright orientation. The angle between the toothpick and the horizontal is the polarization angle.

16. Return to the data collection menu. Follow steps 7–13 using the sunglasses and reflected light instead of the two filters. Start with the toothpick in a horizontal position so that the polarization angle is 0°. Collect data in 15° increments as you did for the two filters. Hold the sunglasses as close as possible to the light sensor as you collect data so that any light entering the sensor must first pass through the sunglasses. Record measurements of background and maximum light intensity in your data table.

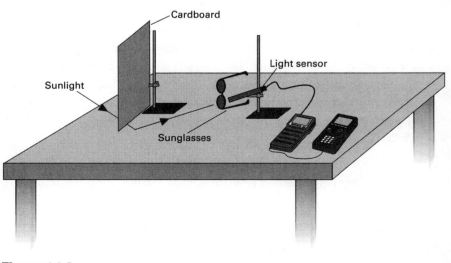

Figure 14-2

DATA TABLE

Artificial light and two filters		Reflected light and sunglasses	
background light reading		background light reading	
maximum intensity		maximum intensity	

ANALYSIS

1. **Comparing and contrasting** Describe the graph of light intensity versus angle that you generated in Part I of the Procedure. Compare the shape of this graph with the graph generated using reflected sunlight and the sunglasses in Part II of the Procedure. How are they the same? How are they different?

2. **Evaluating models** Are both sets of data consistent with Malus's model? Explain any discrepancies from the model that you may have observed.

3. **Investigating differences** Compare the background light intensity and the maximum intensity reading for the two filters and for the sunglasses. Explain the differences you observed.

4. **Calculating percentages** For the sunglasses, the difference between the background and the maximum intensity corresponds to the amount of glare that was eliminated. Calculate the percentage of light intensity reduction for the sunglasses using the following equation:

$$percent\ reduction\ in\ intensity = \frac{maximum\ intensity - background}{maximum\ intensity} \times 100\%$$

CONCLUSIONS

5. **Thinking critically** The background light intensity was probably higher for the sunglasses than for the two filters. Why might the level of background light be important when choosing sunglasses?

6. **Thinking critically** In addition to polarizing the lenses, another way to block glare is to make a pair of sunglasses darker. Why is this not an effective strategy to eliminate glare for the wearer?

7. **Applying the model** Why should a manufacturer of polarized sunglasses take care to place the lenses at a particular orientation in the frames? If a manufacturer wanted to make the claim that a particular brand of polarized glassed eliminated 99 percent of all glare, how many degrees of latitude would they have in installing the lens correctly?

8. **Reaching conclusions** Compare the percentage of light reduction for the sunglasses you tested with the results of others in the classroom. Evaluate the effectiveness of polarized lenses to reduce glare in general.

9. **Presenting conclusions** Polarized lenses cost between 20 percent and 50 percent more than an equivalent non-polarized pair. Write a short consumer report that uses your data to answer the question, "Are polarized sunglasses worth paying an extra cost?"

EXTENSIONS

1. **Extending the model** Set your two filters so that the polarizing axes are at 90°. Add a third polarizing filter between the first two and collect data of the transmitted intensity as a function of the angle of the middle filter. Propose an explanation for your results.

2. **Comparing results** Read a consumer report on sunglasses and compare your results with those in the report.

3. **Extending research** Use your library or Internet resources to research how polarized sunglasses are made. Write a paragraph summarizing the process.

Technology Lab

Capacitors

Designing a timer

A team of engineers is designing a timing device that will allow a strobe to flash for a short, specific amount of time. The engineers have decided to use a combination of a capacitor and a resistor as the basis of the timing device. When a charged capacitor is discharged through a resistor, the capacitor loses its charge in a time-dependent manner. The rate at which the charge is lost is the same each time the capacitor is fully charged and then discharged. Because the potential difference across the capacitor is proportional to the charge on the capacitor, the potential difference across the capacitor drops in the same predictable way as the charge on the capacitor.

The proposed timing device will continuously measure the potential difference across the capacitor. When the potential difference is 90 percent of its maximum, the timing device will turn the strobe on. When the potential drops below 90 percent of the maximum, the strobe is turned off.

The strobe will be designed to flash for 0.01, 0.05, 0.1, and 0.25 seconds. Before they can finish designing the timer, the engineers need to know the discharge rates for some specific capacitor/resistor combinations. They have asked you to help them build and calibrate a prototype circuit and then determine the specific potential values that can be used to turn the strobe on and off.

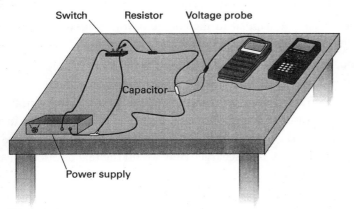

Figure 18-1

OBJECTIVES

- **Model** a timing device using a circuit with a resistor and a capacitor.
- **Measure** the change in potential over time for a circuit containing a capacitor and resistor as the capacitor discharges.
- **Derive** the time constant of a prototype resistor-capacitor circuit.
- **Apply** the results to a strobe flash system.

MATERIALS

✔ graphing calculator with link cable

✔ CBL system

✔ PHYSICS application loaded in calculator

✔ TI or Vernier voltage probe

✔ 10 μF nonpolarized capacitor

✔ 22 and 47 kΩ resistors

✔ 9 V battery with battery clip

✔ single-pole, double-throw switch

✔ connecting wires and clips

SAFETY

- Never close a circuit until it has been approved by your teacher. Never rewire or adjust any element of a closed circuit.

- Never work with electricity near water.

- Do not work with any batteries, electrical devices, or magnets other than those provided by your teacher.

DEVELOPING THE MODEL

As a capacitor discharges, the charge on the capacitor is depleted, reducing the potential difference across the capacitor. The reduction in potential difference reduces the current. This process results in a current that decreases exponentially over time.

The engineers will put capacitors and resistors into a circuit such as the one shown in **Figure 18-2**. After the capacitor is charged to its maximum potential, V_0, the capacitor will be discharged. In this activity, you will monitor the change in potential difference across the capacitor over time. You will then fit an exponential model to the data and use the model to predict the correct voltages for turning the strobe light on and off.

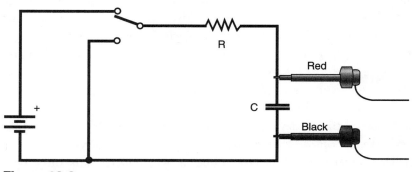

Figure 18-2

Before starting this activity, answer the following questions:

1. **Figure 18-2** shows the prototype timer circuit that you will use to collect data. Notice the position of the resistor and the capacitor. With the switch in the present position, will the capacitor be charged? Explain your answer.

2. Describe what will happen to the charge on the capacitor in **Figure 18-2** when the switch is moved to the other position.

3. The equation $\Delta V_t = \Delta V_0 e^{-\frac{t}{RC}}$ can be used to model the change in the potential difference over time. ΔV_0 is the initial potential difference across the capacitor, ΔV_t is the instantaneous potential difference with respect to time, t is time, R is the resistance in ohms, and C is the capacitance in farads. According to this equation, how will an increase in the capacitance of the circuit affect the rate of discharge? How would the rate of discharge change if the resistance were increased? If you want the timer to measure extremely short periods of time, would you want to use a large resistor or a small resistor? Explain why.

4. The value RC (the circuit resistance multiplied by the total capacitance) is known as the time constant of the circuit. What is the relationship between the time constant and the rate of discharge?

5. Why would it be important to collect calibration data using a prototype circuit that is as close as possible to the one that will actually be used?

PROCEDURE

1. Build the circuit for the prototype timer as shown in **Figure 18-2** with the 10 μF capacitor and the 47 kΩ resistor. Record the values of your resistor

and capacitor in a data table like the one shown in the Data Table section. Also record any tolerance values marked on the resistor and the capacitor.

2. Connect the voltage probe to the CH1 input on the CBL. Then connect the probe leads across the capacitor, with the red (positive lead) connected to the side of the capacitor that is connected to the resistor. Connect the black lead to the other side of the capacitor. Have your teacher approve your circuit before you proceed.

3. Turn on the CBL unit and the calculator. Start the PHYSICS application and proceed to the MAIN MENU.

4. Set up the calculator and CBL for the voltage probe.
   - Select SET UP PROBES from the MAIN MENU.
   - Select ONE as the number of probes.
   - Select VOLTAGE from the SELECT PROBE menu.
   - Confirm that the voltage probe is connected to CH 1, and press [ENTER].

5. Monitor the input to determine the maximum voltage your battery produces.
   - Select COLLECT DATA from the MAIN MENU.
   - Select MONITOR INPUT from the DATA COLLECTION menu.

6. Charge the capacitor for 10 seconds with the switch in the position shown in **Figure 18-2.** Watch the reading on the calculator screen and note the maximum value reached. You will need this value for the next step. Press [+] to leave the monitor mode and select RETURN TO MAIN from the DATA COLLECTION menu.

7. In the timing device, the strobe will be turned on when ΔV_t is at 90 percent of ΔV_0. Set up the calculator and CBL for triggering the start of data collection at this point.
   - Select TRIGGERING from the MAIN MENU.
   - Select CHANNEL 1 from the TRIGGERING menu.
   - Select DECREASING from the TRIGGER TYPE menu.
   - For the trigger threshold, enter 90 percent of the maximum voltage you observed in step 6. Note that you can enter an expression at the calculator's prompt for a trigger level. That is, you can press 0.9 [×] (*maximum voltage*) [ENTER] where *maximum voltage* is the maximum voltage observed in step 6.
   - Enter "0" for PRESTORE IN PERCENT.

8. Set up the calculator and CBL for data collection.
   - Select COLLECT DATA from the MAIN MENU.
   - Select TIME GRAPH from the DATA COLLECTION menu.
   - Enter "0.01" as the time between samples in seconds.
   - Enter "50" as the number of samples.
   - Press [ENTER], and select USE TIME SETUP to continue. Note: If you want to change the sample time or sample number you entered, select MODIFY SETUP.

9. Verify that the switch has been in the position illustrated in **Figure 18-2** for ten seconds, ensuring that the capacitor is charged.

10. Press [ENTER] to begin data collection. Wait a moment, then move the switch to the other position. This will discharge the capacitor. The CBL will not begin collecting data until the measured voltage reaches the trigger value.

11. Press [ENTER], and note the shape of your graph of potential difference versus time.

12. Press [ENTER], and select NO to return to the MAIN MENU.

13. Next, fit the exponential function $y = Ae^{-B*x}$ to your data.

   • Select ANALYZE from the MAIN MENU.

   • Select CURVE FIT from the ANALYZE MENU.

   • Select EXPONENT L_1, L_2 from the CURVE FIT menu.

   • Record the value of the parameters A and B in your data table.

   • Press [ENTER] to view the fitted equation along with your data.

14. Sketch the graph of potential difference plotted against time.

15. Repeat this experiment two more times using the 47 kΩ resistor. Record all data in your data table.

16. Rebuild your circuit using the 22 kΩ resistor and repeat steps 6–14 three more times. Have your teacher approve your circuit before you collect data. Record all data in your data table.

DATA TABLE

Resistance, $R(k\Omega)$	Capacitance, $C(\mu F)$	Time constant, $RC(s)$	Curve fit parameters			Average of $1/B$
			A	B	$1/B$	
47	10					
47	10					
47	10					
22	10					
22	10					
22	10					

ANALYSIS

1. **Calculating** Calculate the time constant $(R \times C)$ for both prototype circuits based on the labeled resistance and capacitance. Record your answer in your data table.

2. **Interpreting models** What does the parameter A in the exponential model actually represent?

3. **Interpreting models** What does the parameter B in the exponential model actually represent?

4. **Calculating** For each trial, calculate the value of $1/B$ and enter it in your data table.

5. **Calculating averages** Calculate an average value of $1/B$ for each of the two resistors you used in this experiment.

6. **Comparing results** Compare the average value of $1/B$ with the time constant you calculated based on the labeled resistance and capacitance. Are these two values within the allotted 5 percent tolerances claimed by the manufacturer? Explain.

7. **Identifying relationships** When you changed the value of the resistor in the circuit, what was the effect on the way the capacitor discharged?

CONCLUSIONS

8. **Reaching conclusions** Based on your repeated trials, how much variation is there in the prototype components? Would you be able to use this circuit in a reliable timing device?

9. **Applying results** For each of the required strobe flash times (0.01, 0.05, 0.1 and 0.5 seconds) state the appropriate resistor that should be used. Also determine the appropriate ΔV_0 and ΔV_t that should be used to turn the strobe on and off. Present your results in a table.

EXTENSIONS

1. **Graphing** Make a plot of $\ln(\Delta V_t)$ versus time for the capacitor discharge. What is the meaning of the slope of this plot? How is it related to the time constant, RC?

2. **Analyzing results** What percentage of the initial potential difference remains after one time constant has passed, after two time constants have passed, and after three time constants have passed?

3. **Applying results** Instead of a resistor, use a small flashlight bulb in the circuit. To light the bulb for a perceptible time, use a large capacitor (approximately 1 F). Collect data, and explain the shape of the graph.

4. **Extending the model** Try using resistors and capacitors with different values to see how the capacitor discharge curves change. Have your teacher approve your circuits before you collect data.

5. **Making predictions** Try using two 10 μF capacitors wired in parallel. Predict what will happen to the time constant. Have your teacher approve your circuit before you proceed. Repeat the discharge measurement and perform a curve fit to determine the time constant of the new circuit.

6. **Making predictions** Try using two 10 μF capacitors wired in series. Predict what will happen to the time constant. Have your teacher approve your circuit before you proceed. Repeat the discharge measurement and determine the time constant for the new circuit using a curve fit.

HOLT PHYSICS
Technology Lab

Electrical Energy

OBJECTIVES

- **Model** a toy car and evaluate the efficiency of an electric motor under different loads.

- **Measure** the power and electrical energy used by an electric motor under different loads.

- **Calculate** the efficiency of the motor using the data collected.

- **Evaluate** the motor efficiency and make recommendations regarding design modifications.

MATERIALS

✔ graphing calculator with link cable

✔ CBL system

✔ PHYSICS application loaded in calculator

✔ Vernier current and voltage probe system

✔ 2 Vernier CBL adapter cables

✔ adjustable-voltage DC power supply

✔ electric motor

✔ mass set

✔ string

✔ table clamp

✔ utility clamp

✔ short dowel rod or thread spool

✔ wires

✔ clips to hold wires

The efficiency of an electric motor

AutoRemote is a company that builds remote-controlled toys such as cars, boats, and planes. Each toy uses a bank of eight 1.5 V batteries to power a 12 V DC motor. AutoRemote's most popular remote control car uses eight rechargeable batteries that provide about 15 minutes of play before needing to be recharged. Recently several new competitors have introduced remote control automobiles. When these new cars were tested by an independent consumer group, data showed that the cars produced by the newer companies could run for up to 25 minutes before their batteries needed to be recharged.

The design staff of AutoRemote has decided to consider redesigning their toy. They are focusing on two areas in the redesign process: the overall weight of the toy, and the efficiency of the electric motor. They have asked you to collect data that will address two questions:

(1) What is the overall operating efficiency of the electric motor?

(2) How does the load, the toy's weight, affect the efficiency of the motor?

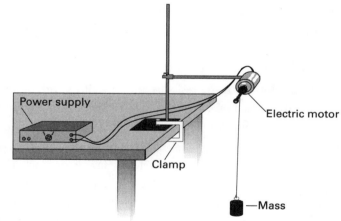

Figure 19-1

SAFETY

- Never close a circuit until it has been approved by your teacher. Never rewire or adjust any element of a closed circuit.

- Never work with electricity near water; be sure the floor and all work surfaces are dry.

- Do not work with any batteries, electrical devices, or magnets other than those provided by your teacher

- Perform this experiment in a clear area. Attach masses securely. Falling, dropped, or swinging objects can cause serious injury.

DEVELOPING THE MODEL

Figure 19-1 shows an experimental setup in which an electric motor is lifting a small mass against the force of gravity. In this situation, the mechanical work performed by the motor is equal to the change in gravitational potential energy of the mass. If you use a current and voltage probe system to measure the current in the motor and the potential difference across the motor as it performs the work, you will be able to calculate the power and the energy used by the motor during the task. You can then compare this value for the energy with the increase in gravitational potential energy of the mass to determine the efficiency of the motor. Changing the mass changes the load on the motor. By performing the experiment with different masses, you can investigate the efficiency of the motor under different loads.

Before starting this activity, answer the following questions:

1. What measurements must be made during the experiment to determine the increase in the potential energy of the mass being lifted by the motor?

2. If you know the current in the motor and the potential difference across the motor at any moment in time, you can calculate power. What is the mathematical relationship between potential difference, current, and power?

3. When power is multiplied by time, the units are W·s. How does this relate to the units for energy? Explain.

4. Would it be possible to determine the efficiency of the motor using this model if the motor was allowed to run with no mass attached? Why or why not?

5. Predict how the work performed by the motor would be affected if a very large load were placed on the motor. How would this affect the calculated efficiency?

6. Set up the motor apparatus as shown in **Figure 19-1,** and connect the motor to a power supply. Do not connect the voltage and current probe at this point. Find the axle on the motor and make sure the string will wind around the axle as the mass is lifted. Allow room for the mass to be lifted at least 0.5 m. Make a loop at the end of the string and attach a 10 g mass. Make sure all wires are out of the way of the motor and the mass. Set the voltage control on your power supply to 0 V. Turn on the power supply and gradually increase the voltage setting. Watch the motor to see when it starts to turn. Control the voltage so that the motor turns and lifts the mass slowly. Set the control to 0 V when the mass reaches the top. Describe the energy changes that take place as the mass is lifted.

7. What would you expect the efficiency of this electric motor to be? That is, what percentage of the electrical energy used by the motor goes into lifting the mass?

PROCEDURE

1. Connect the DIN 1 plug from the dual-channel amplifier to CH 1 on the CBL unit. Connect the DIN 2 plug to CH 2. Connect a voltage probe to PROBE 1 on the dual-channel amplifier. Connect a current probe to PROBE 2. Use the black link cable to connect the CBL unit to the calculator. Firmly press in the cable ends.

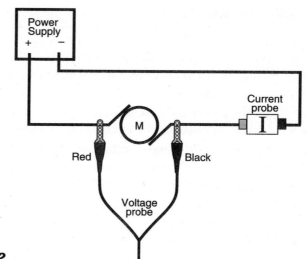

Figure 19-2

2. Connect the circuit as shown in **Figure 19-2.** Make sure that the positive (red) lead of the voltage probe and the red terminal from the current probe are oriented toward the positive terminal as shown. Before you proceed, have your teacher approve your circuit.

3. Turn on the CBL unit and the calculator. Start the PHYSICS application and proceed to the MAIN MENU.

4. Set up the calculator and CBL for the current and voltage probe.

- Select SET UP PROBES from the MAIN MENU.
- Select TWO as the number of probes.
- Select C-V VOLTAGE from the SELECT PROBE menu.
- Confirm that the DIN 1 cord is connected to CHANNEL 1, and the voltage probe is connected to PROBE 1. Press [ENTER].
- Select USE STORED from the CALIBRATION menu.
- Select C-V CURRENT from the SELECT PROBE menu.
- Confirm that the DIN 2 cord is connected to CHANNEL 2, and the current probe is connected to PROBE 2. Press [ENTER].
- Select USE STORED from the CALIBRATION menu.

5. With the power supply turned off, zero the sensors.

- Select ZERO PROBES from the MAIN MENU.
- Select ALL CHANNELS from the SELECT CHANNEL menu.
- Follow the instructions on the calculator screen to zero the sensors.

6. The procedures in this lab call for a 12 V motor; however, your classroom may have different motors. **Check the voltage rating of your motor. This is the maximum voltage you should use.** If you are not sure of the rating, ask your instructor. Record the voltage rating in a data table like the first one shown in the Data Tables section.

7. Set up the calculator and CBL for data collection.

- Select COLLECT DATA from the MAIN MENU.
- Select TIME GRAPH from the DATA COLLECTION menu.
- Enter "0.25" as the time between samples, in seconds.

- Enter "99" as the number of samples (the CBL will collect data for about 25 seconds).

- Press ENTER, then select USE TIME SETUP to continue. If you want to change the sample time or sample number, select MODIFY SETUP instead.

8. Check to make sure that the 10 g mass is securely attached to the string. Use a piece of tape to mark the starting position of the mass. You will use the same starting position for each trial. Measure the distance the mass will rise, and record the value in your data table.

9. You should be ready to measure the current in the motor and the potential difference across the motor as it lifts the mass.

- Set the voltage control on your power supply to 0 V and turn it on.

- Press ENTER to begin data collection.

- Gradually increase the voltage setting on the power supply. Watch the motor to see when it starts to turn slowly. Control the voltage so that the motor turns and lifts the mass slowly. The CBL will collect data for about 25 seconds, so you should make sure the motor will be able to make the lift in this time.

- Set the control to 0 V when the mass reaches the top. This must occur before data collection ends.

- Press ENTER to see the graph of voltage versus time; press ENTER again to see the graph of current versus time.

Check the graph to make sure that you have voltage and current readings for the entire time the mass was being lifted. If necessary, adjust your setup and repeat this step until you get a good run. To repeat, press ENTER and select YES from the REPEAT? menu.

10. Calculate the moment-by-moment power from the product of the voltage data (stored in the list L_2 in your calculator) and the current data (stored in the list L_3). This calculation requires working with data lists outside of the PHYSICS application, so press ENTER, select NO from the REPEAT menu, and select QUIT from the MAIN MENU.

- Press [2nd] L2 [×] [2nd] L3 [STO▶] [2nd] L6. This will store the power data in the list L6.

11. Next determine the total energy consumed while the motor lifted the mass. The energy used is the product of the rate at which it was used and the time over which is was used. Because your data was taken in discrete time intervals, the total energy used is equivalent to the sum of the power at each moment multiplied by the length of each time interval (0.25 s).

- From the calculator home screen, press [2nd] LIST.

- Use the [▶] key to move to the MATH menu.

- Press "5" to select sum. Note that the calculator has already entered the left parenthesis for you.

- Press [2nd] L6 [×] 0.25 [)] and press ENTER.

- The result is the energy consumed by the motor as it lifted the mass. Record this value in a data table like the second one shown in the Data Tables section.

12. Restart the PHYSICS application and proceed to the MAIN MENU.

13. Increase the load by 10 g and repeat steps 7–12. Again, make sure you get voltage and current data for the entire lift. For each trial, note the mass lifted and energy used in your data table.

14. Repeat step 13 five more times, or until the motor will not lift the load without exceeding its voltage rating. Be sure that the mass is lifted the same distance each time.

DATA TABLES

Distance mass is lifted (m)	
Voltage rating of motor (V)	

Trial	Load lifted (kg)	Force exerted (N)	Electrical energy used (J)	Mechanical energy output (J)	Efficiency (%)
1					
2					
3					
4					
5					
6					
7					

ANALYSIS

1. Calculating Calculate the force exerted to overcome the force of gravity for each load ($F = mg$). Enter the value in your data table.

2. Calculating For each trial, calculate the increase in gravitational potential energy of the mass. The increase in gravitational potential energy is equal to the mechanical energy output of the motor. Record the values in your data table.

3. Calculating For each trial, calculate the efficiency of the motor using the following equation:

$$efficiency = \frac{mechanical\ energy\ output}{electrical\ energy\ input} \times 100\%$$

Record your answer in your data table.

4. Analyzing results When you were adding masses, was there a load at which the efficiency became 0? Explain how this could happen.

CONCLUSIONS

5. Applying ideas In each trial, the work done by the motor was less than the electrical energy used by the motor. Besides being used to lift the mass, what happened to the electrical energy that went to the motor?

6. **Drawing conclusions** What is the highest efficiency that AutoRemote can expect from this motor?

7. **Evaluating data** If drag is defined as the total force that the motor must overcome when propelling a car, what drag would result in the most efficient use of the motor? What drag would result in the longest battery life?

8. **Designing models** Suggest some design changes that can be incorporated into the car so that it can operate over a range of speeds for a longer period of time before the batteries need recharging. Would these design changes increase the cost of the toy? Explain.

EXTENSIONS

1. **Graphing** On graph paper or your graphing calculator, plot a graph of the efficiency of the motor as a function of the load.

2. **Designing experiments** Design an experiment to investigate the efficiency of the motor at different speeds using the same load. If you have time and your teacher approves, perform your experiment.

3. **Extending research** Research motor design to determine how design changes may affect efficiency.

4. **Calculating** Calculate the average power generated by the motor under different loads. Convert the value to units of horsepower.

5. **Applying ideas** Can you use your motor as a generator? Raise the mass using the motor and hold it at the top by hand. Turn off the power supply and remove it from the circuit. Replace the power supply with a 10 Ω resistor, and have your teacher approve your circuit. Obtain current and voltage data to find the power generated by the falling mass as it turns the motor. You may need to increase the mass. Compare the power generated to the change in gravitational potential energy of the falling mass.

Technology Lab

Series and Parallel Circuits

OBJECTIVES

- **Develop** a model for measuring currents in both parallel and series circuits.

- **Measure and analyze** voltage and current in standard series and parallel circuits designed with two resistors.

- **Design** a series of circuits using a combination of series and parallel resistor elements to achieve specific currents.

- **Evaluate** the circuit design strategies of Current Design's new line of instrument pickups and microphones.

MATERIALS

- ✔ graphing calculator with link cable
- ✔ CBL system
- ✔ PHYSICS application loaded in calculator
- ✔ Vernier current and voltage probe system
- ✔ two Vernier adapter cables
- ✔ low-voltage DC power supply
- ✔ two 50 Ω resistors
- ✔ two 68 Ω resistors
- ✔ momentary-contact switch
- ✔ connecting wires or breadboard

Designing Amplifier Circuits

Current Designs, Inc., is in the process of developing a new line of musical instrument pickups and microphones that will allow acoustic instruments, such as a guitar, to be amplified electronically. Each of the pickups in the line will be driven by a small preamplifier to boost the sound before it reaches the main amplifier. The preamplifier is powered by two 1.5 V batteries.

Circuit designs for the various products call for currents of 0.013 A, 0.017 A, 0.030 A, and 0.208 A, with a 5 percent tolerance in each case. Current Designs is hoping to keep costs down by purchasing large quantities of two types of resistors, and using only 50 Ω resistors and 68 Ω resistors in all circuits. The engineer has provided you with a set of these resistors. Your task is to determine whether these resistors can be wired (either in parallel, series, or a combination of the two) to provide the desired currents.

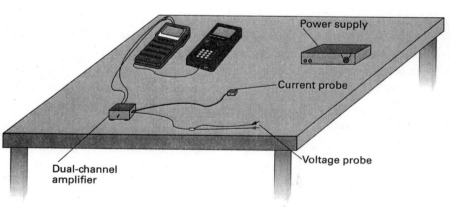

Figure 20-1

SAFETY

- Never close a circuit until it has been approved by your teacher. Never rewire or adjust any element of a closed circuit.

- Never work with electricity near water; be sure the floor and all work surfaces are dry.

- Do not work with any batteries, electrical devices, or magnets other than those provided by your teacher.

DEVELOPING THE MODEL

The ability to produce several different products with a small number of parts helps a manufacturer to cut production costs. Before you design the circuits needed by your company, you will first build several simple series and parallel circuits. By monitoring these circuits using a current and voltage probe system, you can see how different arrangements of resistors will affect the current. Once you understand how circuit design affects the current, you will be able to design several circuits that have specific currents.

1. Manufacturers of resistors use a color code to indicate a tolerance rating. Tolerance is a percentage rating, showing how much the actual resistance could vary from the labeled value. Determine the tolerance ratings for the 50 Ω resistor and the 68 Ω resistor you are using. Calculate the range of resistance values that fall within the tolerance range for each resistor.

2. Suppose that a circuit contained a single light bulb. What would happen to the current in the circuit if a second or third light bulb were added in series? How would this affect the brightness of the bulbs? Explain. What would happen to the circuit containing a single light bulb if a second or third bulb were added in parallel to the first? How would this affect the brightness of the bulbs? Explain.

3. How can you determine the resistance of a light bulb or other circuit element if you are only able to measure current and voltage?

4. Even though it is possible to calculate the current for an unknown element, explain why it would be important to measure the current for a proposed circuit design using the actual components you will be using to build the circuit.

PROCEDURE

Part I Series Circuits

1. Connect DIN 1 on the dual-channel amplifier to CH 1 on the CBL unit. Connect DIN 2 TO CH 2. Connect a current probe to PROBE 1 on the dual channel amplifier. Connect a voltage probe to PROBE 2. Use the black link cable to connect the CBL unit to the calculator. Firmly press in the cable ends.

2. Turn on the CBL unit and the calculator. Start the PHYSICS application and proceed to the MAIN MENU.

3. Set up the calculator and CBL for the current and voltage probe.
   - Select SET UP PROBES from the MAIN MENU.
   - Select TWO as the number of probes.
   - Select C–V CURRENT from the SELECT PROBE menu.
   - Confirm that the DIN 1 cable from the dual channel amplifier is connected to CHANNEL 1, and press [ENTER].
   - Select USE STORED from the CALIBRATION menu.
   - Select C–V VOLTAGE from the SELECT PROBE menu.
   - Confirm that the DIN 2 cable from the dual channel amplifier is connected to CHANNEL 2, and press [ENTER].
   - Select USE STORED from the CALIBRATION menu.

4. Now you will zero both probes with no current and with no voltage applied. Connect the black and red clips together for this step only.

- Select ZERO PROBES from the MAIN MENU.
- Select ALL CHANNELS from the SELECT CHANNEL menu.
- Press TRIGGER on the CBL unit.

5. Monitor the current and voltage readings.

- Select COLLECT DATA from the MAIN MENU.
- Select MONITOR INPUT from the DATA COLLECTION menu.
- The calculator screen now shows the current (CH 1) and potential (CH 2), updated about once a second.

6. If you have an adjustable power supply, set it at 3.0 V.

7. Connect the series circuit shown in **Figure 20-2** using 50 Ω resistors for R_1 and R_2. Notice the voltage probe is used to measure the voltage applied to both resistors. The red terminal of the current probe should be toward the positive terminal of the power supply. Have your teacher approve your circuit before you proceed.

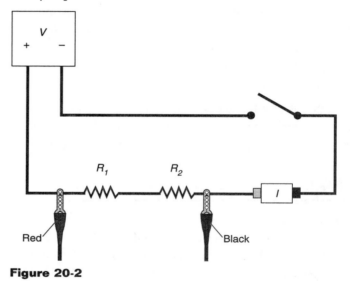

Figure 20-2

8. You can take readings from the calculator screen at any time. To test your circuit, close the switch to complete the circuit; hold for several seconds. Both current and voltage readings should increase. If they do not, open the switch and recheck your circuit.

9. Close the switch to complete the circuit again and read the current (I) and total voltage (ΔV_{tot}). Record the values in a data table like the one shown for Part I in the Data Tables section.

10. Connect the leads of the voltage probe across R_1. Close the switch to complete the circuit and read this voltage (ΔV_1). Record this value in your data table.

11. Connect the leads of the voltage probe across R_2. Close the switch to complete the circuit and read this voltage (ΔV_2). Record this value in your data table.

12. Repeat steps 8–11 with a 68 Ω resistor substituted for R_2.

13. Repeat steps 8–11 with a 68 Ω resistor used for both R_1 and R_2.

Part II Parallel Circuits

14. Connect the parallel circuit shown in **Figure 20-3** using 50 Ω resistors for both R_1 and R_2. As in the previous circuit, the voltage probe is used to measure the voltage applied to both resistors. The red terminal of the current probe should be toward the positive terminal of the power supply. The current probe is used to measure the total current in the circuit. Have your teacher approve your circuit before you proceed.

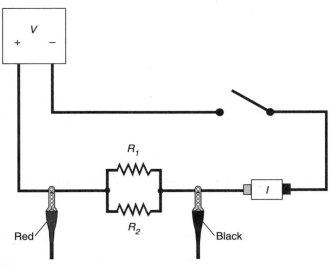

Figure 20-3

15. As in Part I, you can take readings from the calculator screen at any time. To test your circuit, close the switch to complete the circuit; hold for several seconds. Both current and voltage readings should increase. If they do not, open the switch and recheck your circuit.

16. Close the switch to complete the circuit again and read the total current (I) and total voltage (ΔV_{tot}). Record the values in a data table like the one shown for Part II in the Data Tables section.

17. Connect the leads of the voltage probe across R_1. Close the switch to complete the circuit and read the voltage (ΔV_1) across R_1. Record this value in your data table.

18. Connect the leads of the voltage probe across R_2. Close the switch to complete the circuit and read the voltage (ΔV_2) across R_2. Record this value in your data table.

19. Repeat steps 16–18 with a 68 Ω resistor substituted for R_2.

20. Repeat steps 16–18 with a 68 Ω resistor used for both R_1 and R_2.

Part III Designing Circuits

21. Study your data from Part I and Part II to determine how resistors can affect the current in series and parallel circuits. Make a preliminary design of a circuit that will result in a current of 0.017 A. You may use from one to four resistors in your design. In addition, you may use a simple series or parallel circuit, or you may design a circuit that combines resistors in both series and parallel configurations. Make a circuit diagram and write a short description of the circuit in a table like the one shown for Part III in the Data Tables section.

22. Set up the circuit you designed in step 21 and have your teacher approve your circuit. Then repeat steps 16–18 to test the current in your circuit. Record the current measurements in your data table.

23. If your current was within the required 5 percent tolerance specified by the engineer, move on to design additional circuits with the specified currents. If your current was not within the required range, revise and retest your circuit design until your circuit meets the requirements.

24. Continue until you have successfully designed all four required circuits. Record your final current measurements in your data table.

DATA TABLES

Part I

R_1 (Ω)	R_2 (Ω)	I (A)	ΔV_1 (V)	ΔV_2 (V)	R_{eq} (Ω)	ΔV_{tot} (V)
50	50					
50	68					
68	68					

Part I

R_1 (Ω)	R_2 (Ω)	I (A)	ΔV_1 (V)	ΔV_2 (V)	R_{eq} (Ω)	ΔV_{tot} (V)
50	50					
50	68					
68	68					

Part III

Circuit	Desired current (A)	Circuit description	Measured current (A)	percent deviation (%)
1	0.017			
2	0.013			
3	0.032			
4	0.330			

ANALYSIS

1. **Identifying relationships** Examine the results of Part I. What is the relationship between the three voltage readings: ΔV_1, ΔV_2, and ΔV_{tot}?

2. **Calculating** Using the measurements you have made above and your knowledge of Ohm's law, calculate the equivalent resistance (R_{eq}) of the circuit for each of the three series circuits you tested in Part I. Describe a general rule for the equivalent resistance (R_{eq}) of a series circuit that has two resistors.

3. **Calculating** Using your measurements and your knowledge of Ohm's law, calculate the equivalent resistance (R_{eq}) of the circuit for each of the three parallel circuits you tested in Part II. Describe a general rule for the equivalent resistance of a parallel circuit that has two resistors.

4. **Interpreting results** Using your data from Part I, describe what happens to current as you add more resistors in a series circuit.

5. **Interpreting results** Using your data from Part II, describe what happens to current as you add more resistors in a parallel circuit.

6. **Evaluating results** For each of the circuits you designed in Part III, calculate the percent deviation from the target resistance using the following formula:

$$percent\ deviation = \frac{target\ current - measured\ current}{target\ current} \times 100\%$$

Record your results in your data table for Part III.

CONCLUSIONS

7. **Evaluating models** For each of the three series circuits, compare the experimental results with the value you calculated for the resistance. Were the measured current values within range of the tolerances specified on the resistors? Explain.

8. **Making predictions** If it were not possible to add more resistance, what other changes could be made to lower the current?

9. **Reaching conclusions** Would the circuits you designed be acceptable for use in the pickups and microphones described in the introduction? Why or why not?

10. **Reaching conclusions** Would it be possible to design a circuit having a current of less than 0.001 A using only 50 Ω and 68 Ω resistors? What about a circuit having a current greater than 10 A? Are there lower or upper current limits to circuits that are designed using any possible combination or quantity of 50 Ω and 68 Ω resistors? Explain.

EXTENSIONS

1. **Calculating** Calculate the R_{eq} for the circuits you designed in Part III of this lab based on the values of the individual resistors. Compare these with your measurements.

2. **Designing experiments** Design an experiment to determine the resistance of several different kinds and gauges of wires. Include copper, aluminum, and nichrome wire. If you have time and your teacher approves of your plan, perform this experiment.

3. **Evaluating results** Repeat Parts I and II of this experiment using two different small electric light bulbs in place of resistors. Because the light bulbs are not rated for resistance, you will have to rely on current and voltage measurements. In addition to collecting data using the CBL system, evaluate the brightness of the bulbs qualitatively or with a light sensor.

Technology Lab

Magnetic Field Strength

OBJECTIVES

- **Model** a current meter that uses a magnetic field sensor to measure current.

- **Investigate** the relationship between the magnetic field and the number of turns in a coil.

- **Investigate** the relationship between the magnetic field and the current in a coil.

- **Determine** the current in a low-voltage DC appliance using your model.

MATERIALS

- ✔ graphing calculator with link cable
- ✔ CBL system
- ✔ PHYSICS application loaded in calculator
- ✔ Vernier magnetic field sensor and CBL adapter cable
- ✔ TI or Vernier voltage probe
- ✔ 1 Ω power resistor
- ✔ long spool of insulated wire (at least 12 m)
- ✔ low-voltage DC motor, light bulb, or buzzer
- ✔ magnetic compass
- ✔ momentary-contact switch
- ✔ small square or circular frame or shoe box

Building a current meter

A friend of yours has recently built a small vacation cabin in the desert. Because there are steady winds all year, she has decided to design and build a wind-driven generator to supply the cabin with 12 V DC power. She has decided to build all of the components herself, including a simple current meter that will allow her to measure the current at any point in time.

The design of her current meter calls for a magnetic field sensor to be placed inside a coil of wire in which there is household current. When there is current in the coil of wire, a magnetic field is induced, which is measurable by the sensor. The greater the current in the coil, the greater the reading on the magnetic sensor. In order for the current meter to be useful, it must be calibrated so that the reading on the sensor can be converted from milliteslas (mT) to amperes (A). Your friend has asked you to help her refine and calibrate the meter.

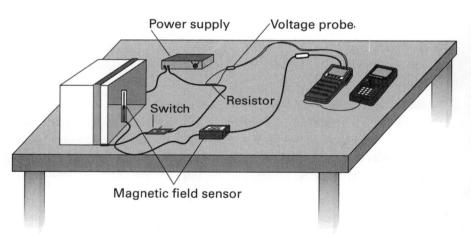

Figure 21-1

SAFETY

- Never close a circuit until it has been approved by your teacher. Never rewire or adjust any element of a closed circuit.

- Never work with electricity near water; be sure the floor and all work surfaces are dry.

- Do not work with any batteries, electrical devices, or magnets other than those provided by your teacher

DEVELOPING THE MODEL

First, you will explore how the magnetic sensor works by measuring the intensity of Earth's background magnetic field. Then you will investigate the relationship between the number of coils and the strength of the magnetic field. Finally, you will calibrate the meter and use it to determine the current in a small DC electrical appliance.

Before starting this activity, answer the following questions:

1. If you are given a voltmeter and a 1 Ω resistor, how can you calculate the current in a circuit?

2. To return the maximum reading, the white dot on the magnetic field sensor should be pointed in the direction of the north pole of the magnetic field. How can you make sure that the sensor is correctly positioned?

3. If you were to graph the current in a coil versus the magnetic field strength of the coil, you would find that the points generally fall on a straight line. How could you use the strength of the magnetic field and the slope of the line to determine the current in a circuit?

4. Before you run current through your circuit, you will zero the sensor so that the background magnetic field will be subtracted from all your readings. Why would it be important to zero the sensor again if you changed your experimental setup in the middle of data collection?

PROCEDURE

Part I Orienting the Sensor

1. Using the long spool of wire, carefully loop the wire ten times around a square or circular frame or box to create a coil of ten turns.

2. Connect the voltage probe to the CH 1 port of the CBL unit. Connect the magnetic field sensor to the CH 2 port of the CBL unit. Set the switch on the sensor to *High*. Use the black link cable to connect the CBL unit to the calculator. Firmly press in the cable ends.

3. Connect the coil, switch, resistor, and power supply as shown in the setup diagram. Connect the voltage probe across the resistor, with the positive (red) lead on the side of the resistor that is connected to the positive side of the power supply. Have your teacher approve your circuit before you proceed.

4. Turn on the CBL unit and the calculator. Start the PHYSICS application and proceed to the MAIN MENU.

5. Set up the calculator and CBL for the voltage probe and the magnetic field sensor.

   • Select SET UP PROBES from the MAIN MENU.

   • Select TWO as the number of probes.

   • Select VOLTAGE from the SELECT PROBE menu.

   • Confirm that the voltage probe is connected to CHANNEL 1, and press ENTER.

   • Select MAGNETIC FIELD from the SELECT PROBE menu.

   • Confirm that the magnetic field sensor is connected to CHANNEL 2, and press ENTER.

- Select USE STORED from the CALIBRATION menu.
- Select HIGH(MTESLA) from the MG FIELD SETTING menu.

6. For now, leave the power supply off. Monitor the magnetic field:
- Select COLLECT DATA from the MAIN MENU.
- Select MONITOR INPUT from the DATA COLLECTION menu. The magnetic field reading will be displayed on the CH 2 line of the calculator screen. The reading is updated about once per second.

7. Hold the plastic rod containing the magnetic field sensor vertically and far from the coil. Slowly rotate the rod around a vertical axis, and look at the readings. Record your observations in a data table like the one shown for Part I in the Data Tables section. Speculate on what might cause variations in the readings, and record your ideas in your data table.

8. Determine the orientation of the sensor for which the magnetic field is at a maximum, and make a note of the direction that the white dot on the sensor points. Compare the direction of the white dot with the direction of the compass needle. Record your observations in your data table.

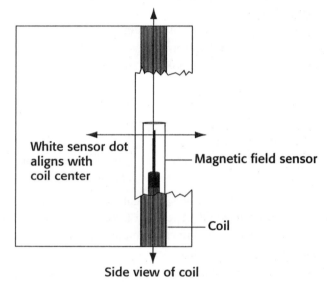

Figure 21-2

9. You now need to determine the direction of the north pole of the magnetic field inside the coil. Close the switch, set the power supply so that the CH 1 display reads 3 V, then open the switch. **Warning: Do not leave the switch closed except when taking measurements.** Leaving the switch closed may cause the coils to overheat and also increases the risk of shock.

10. Place the sensor in a vertical position at the center of the coil, with the white dot facing along the axis of the coil, as shown in Figure 21-2. Close the switch, and make a note of the magnetic field strength. Rotate the magnetic field sensor while you continue to hold the switch closed. Make a note of the position and orientation of the white dot that allows for the maximum reading. Open the switch.

11. Use a support stand and clamp or some other means to hold the magnetic field sensor in the coil in the position and orientation that gives the maximum magnetic field strength reading. The sensor will remain in this position for the rest of the activity.

Part II The Number of Turns and the Magnetic Field Strength

12. Close the switch briefly to make sure that the current is still 3 A (3 V on the CHANNEL 1 display). The current should remain at 3 A throughout Part II.

13. Set up the calculator and CBL for the magnetic field sensor only.

- Remove the voltage probe from CH 1 and move the magnetic field sensor to the CH 1 port of the CBL.
- Select SET UP PROBES from the MAIN MENU.
- Select ONE as the number of probes.
- Select MAGNETIC FIELD from the SELECT PROBE menu.
- Confirm that the magnetic field sensor is connected to CHANNEL 1, and press ENTER.
- Select USE STORED from the CALIBRATION menu.
- Select HIGH(MTESLA) from the MG FIELD SETTING menu.

14. Zero the sensor when there is no current in the coil.

- Verify that the current in the coil is off.
- Select ZERO PROBES from the MAIN MENU.
- Select CHANNEL 1 from the SELECT CHANNEL menu.
- When the reading on the CBL screen is stable, follow the instructions on the calculator screen to zero the sensor.

15. Begin to collect magnetic field data.

- Select COLLECT DATA from the MAIN MENU.
- Select TRIGGER/PROMPT from the DATA COLLECTION menu.
- Close the switch, and press TRIGGER on the CBL to record the magnetic field.
- Open the switch to turn off the current.
- Enter "10" (the number of turns) on the calculator.
- Select MORE DATA from the DATA COLLECTION menu to continue.

16. Make sure the switch is open (off), then remove one loop of wire from the frame to reduce the number of turns by one. Do not shorten the overall length of wire connected to the power supply. Take another data point as in Step 15, entering the appropriate number of turns. If you move the frame or the sensor, make sure that you return it to the same orientation as the previous measurement.

17. Repeat step 16 until you have only one turn of wire left on the frame. After the last data point, instead of selecting MORE DATA, select STOP AND GRAPH from the DATA COLLECTION menu.

18. The graph shows the magnetic field versus the number of turns in the coil. Fit a straight line to the data.

- Press ENTER and select NO to return to the MAIN MENU.
- Select ANALYZE from the MAIN MENU.
- Select CURVE FIT from the ANALYZE MENU.
- Select LINEAR L_1, L_2 from the CURVE FIT menu.
- Record the slope and intercept, with units, in a data table like the one shown for Part II in the Data Tables section.

- Press ENTER to see a graph of your data with the fitted line.
- Sketch your graph, then press ENTER to return to the MAIN MENU.

Part III Calibrating the Meter

19. Carefully wind your coil again so that it is like the first coil in Part II.

20. Repeat steps 3–5 so that you are again collecting data from both the magnetic field sensor and the voltage probe. Make sure the sensor is still positioned for maximum readings.

21. Zero the sensor to subtract the effect of Earth's magnetic field and any local magnetic effects.

- Select ZERO PROBES from the MAIN MENU.
- Select CHANNEL 2 from the SELECT CHANNEL menu.
- Press and release CH VIEW on the CBL until the CH 2 indicator is flashing. This displays the magnetic field reading on the CBL.
- When the reading is stable, press TRIGGER on the CBL unit.

22. Now you are ready to collect magnetic field data as a function of current.

- Select COLLECT DATA from the MAIN MENU.
- Select TRIGGER from the DATA COLLECTION menu.
- Close the switch for the rest of this run.
- Press and release CH VIEW on the CBL until the CH 1 indicator is flashing. This displays the voltage across the 1 Ω resistor (equivalent to the current through the resistor in amperes).
- Verify that the current is still 3 A.
- Press TRIGGER on the CBL to record the current and the magnetic field.
- Select CONTINUE from the TRIGGER menu to continue.

23. Now collect five data points in the following manner:

- Press and release CH VIEW on the CBL until the CH 1 indicator is flashing.
- Decrease the current by 0.5 A.
- Press TRIGGER on the CBL to record the magnetic field.
- Select CONTINUE from the TRIGGER menu, decreasing the current by 0.5 A each time.
- After the 0.5 A point, select STOP from the TRIGGER menu.
- Open the switch to turn off the current in the coils.
- Press ENTER to see a graph of magnetic field versus current.

24. The graph shows the magnetic field versus the current in the coil. Fit a straight line to the data.

- Press ENTER and select NO to return to the MAIN MENU.
- Select ANALYZE from the MAIN MENU.
- Select CURVE FIT from the ANALYZE MENU.
- Select LINEAR L_2, L_3 from the CURVE FIT menu.
- Record the slope and intercept, with units, in a data table like the one shown for Part III in the Data Tables section.

- Press [ENTER] to see a graph of your data with the fitted line.
- Sketch your graph, then press [ENTER] to return to the MAIN MENU.

Part IV Determining the current of a simple circuit

25. Remove the 1 Ω resistor from the circuit and replace it with an electric motor, light bulb, buzzer or other low-voltage DC appliance. Check with your teacher to determine the correct voltage to apply to the circuit. Set the voltage on the power supply accordingly. Have your teacher approve your circuit before you proceed.

26. Now you are ready to determine the current in your circuit.

- Select COLLECT DATA from the MAIN MENU.
- Select MONITOR INPUT from the DATA COLLECTION menu.
- Close the switch, then record the voltage and the magnetic field strength in a data table like the one shown for Part IV in the Data Tables section.
- If you wish to test another device, select MORE DATA from the DATA COLLECTION menu to continue.

DATA TABLES

Part I

Observations — position and orientation of sensor

Part II

Magnetic field versus turn parameters	
A (slope)	
B (intercept)	

Part III

Magnetic field versus current parameters	
A (slope)	
B (intercept)	

Part IV

Device	Voltage (v)	Magnetic field strength (T)	Current (A)	Power (W)

ANALYSIS

1. **Interpreting graphs** Based on your graph of the magnetic field versus the number of turns on the coil (step 18), describe the relationship between the number of turns and the magnetic field at the center of the coil.

2. **Identifying relationships** What is the equation of the line that you fitted to the graph in step 18? What do the two constants in the equation represent? Should the line pass through the origin? Why or why not?

3. **Interpreting graphs** Based on your graph of the magnetic field versus current (from step 24), describe the relationship between the current in the coil and the resulting magnetic field at the center of the coil.

4. **Identifying relationships** What is the equation of the line that you fitted to the graph in step 24? What do the two constants in the equation represent? Should the line pass through the origin? Why or why not?

CONCLUSIONS

5. **Drawing conclusions** A current meter that uses a magnetic field sensor can be made with any number of loops. Discuss the advantages and disadvantages of using very low and very high numbers of loops.

6. **Modeling relationships** Rewrite the equation that describes the relationship between magnetic field strength and current so that current is expressed as a function of magnetic field strength.

7. **Calculating** Use the equation you wrote in item 6 to calculate the current in the device(s) you tested in Part IV.

8. **Calculating** Calculate the power drawn by the device you tested in Part IV. Use the following equation:

$$power = current \times potential\ difference$$

9. **Analyzing methods** Would your current meter need to be recalibrated for a 12 V system? Why or why not?

EXTENSIONS

1. **Designing systems** There are many factors that will affect the strength of a magnetic field produced by a coil, including: the number of coils, the diameter of the coils, and the presence of an iron core. Design a current meter that uses less wire but produces a greater magnetic field than the one you used in this experiment. If your teacher approves, build and test the meter.

2. **Extending investigations** Find out what happens when you rotate the magnetic field sensor horizontally instead of vertically. Hold the magnetic field sensor horizontally and collect data while rotating it smoothly about a horizontal axis. Explain where the maximum and minimum readings occur and where zero or near-zero readings occur. Compare your pattern to the data you collect while rotating about a vertical axis.

3. **Extending research** Earth's magnetosphere is affected by many factors and is constantly changing. Do research to collect data on the causes and effects of variation in the magnetic field surrounding Earth. Summarize your findings in a report.